Calcite

The Mineral with the Most Forms

extraLapis English No. 4

Contributions by
Reinhard Balzer, Sarah Bronko, Michael P. Cooper, Rudolf Duthaler, Stanley Dyl, Michael Gray, Günter Grundmann, Mickey Gunter, Rupert Hochleitner, Terry Huizing, Werner Lieber, Guanghua Liu, Bertold Ottens, R. Peter Richards, Lydie Touret and Marc Wilson

Edited by
Terry Huizing, Miranda Jarnot, Günther Neumeier, R. Peter Richards and Gloria Staebler

Translated by
Howard Heitner, Günther Neumeier, Alfredo Petrov and R. Peter Richards

One of Haüy's original crystal models (see article page 26) *Johannes Keilmann collection*

Photos and Diagrams by
Reinhard Balzer, Torbern Bergman, Ulrich Burchard, Michael P. Cooper, Rock Currier, Diderot & d'Alembert, Kevin Downey, Walter Gabriel, Maximilian Glas, Victor Goldschmidt, André Nicolas van Gorp, Lindsay Greenbank, Mickey Gunter, Waltraud Harre, Hans Hauswaldt, René Just Haüy, Rupert Hochleitner, Terry Huizing, Ted Johnson, Ray Lasmanis, Werner Lieber, Alain Martaud, Olaf Medenbach, Hartmut Meyer, Henry Miers, Erich Offermann, Günther Neumeier, Stefan Oeldenberger, Dane Penland, R. Peter Richards, Paul Rustemeyer, Jeffrey Scovil, Steve Smale, Gloria Staebler, Peter Stemvers, Hugo Strunz, Marcel Vanek, Urs Widmer, Frederick Wilda and Debra Wilson

---

IN COLLABORATION WITH LAPIS MAGAZINE,
CHRISTIAN WEISE VERLAG, GERMANY AND LAPIS INTERNATIONAL, USA

Table of Contents: History ◊ Habits

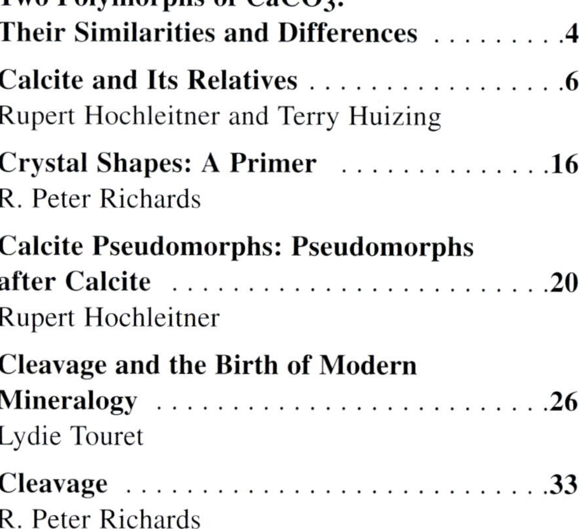

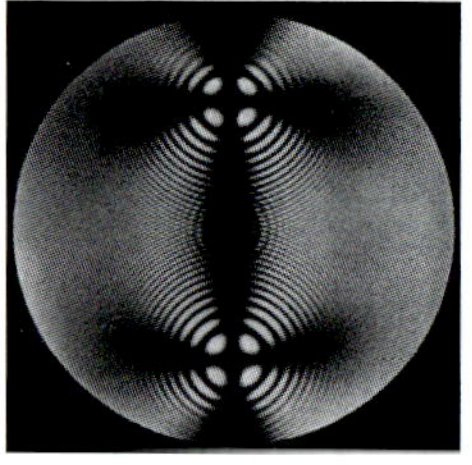

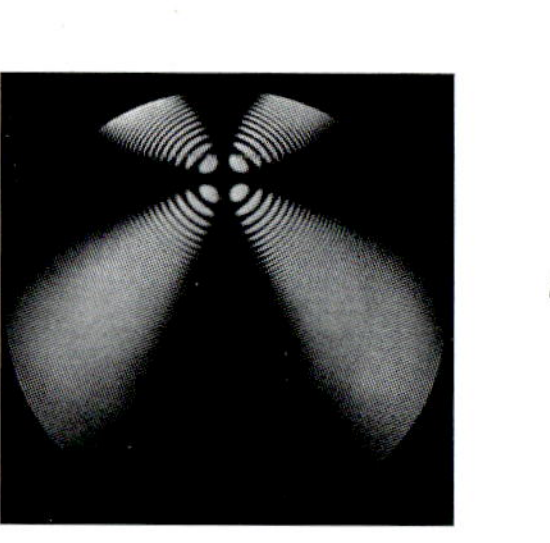

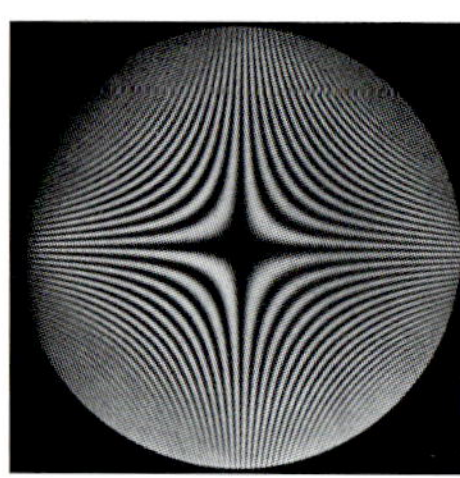

Above: *Four of the more than 1,000 interference patterns prepared and photographed by Hans Hauswalt between 1897 and 1901*

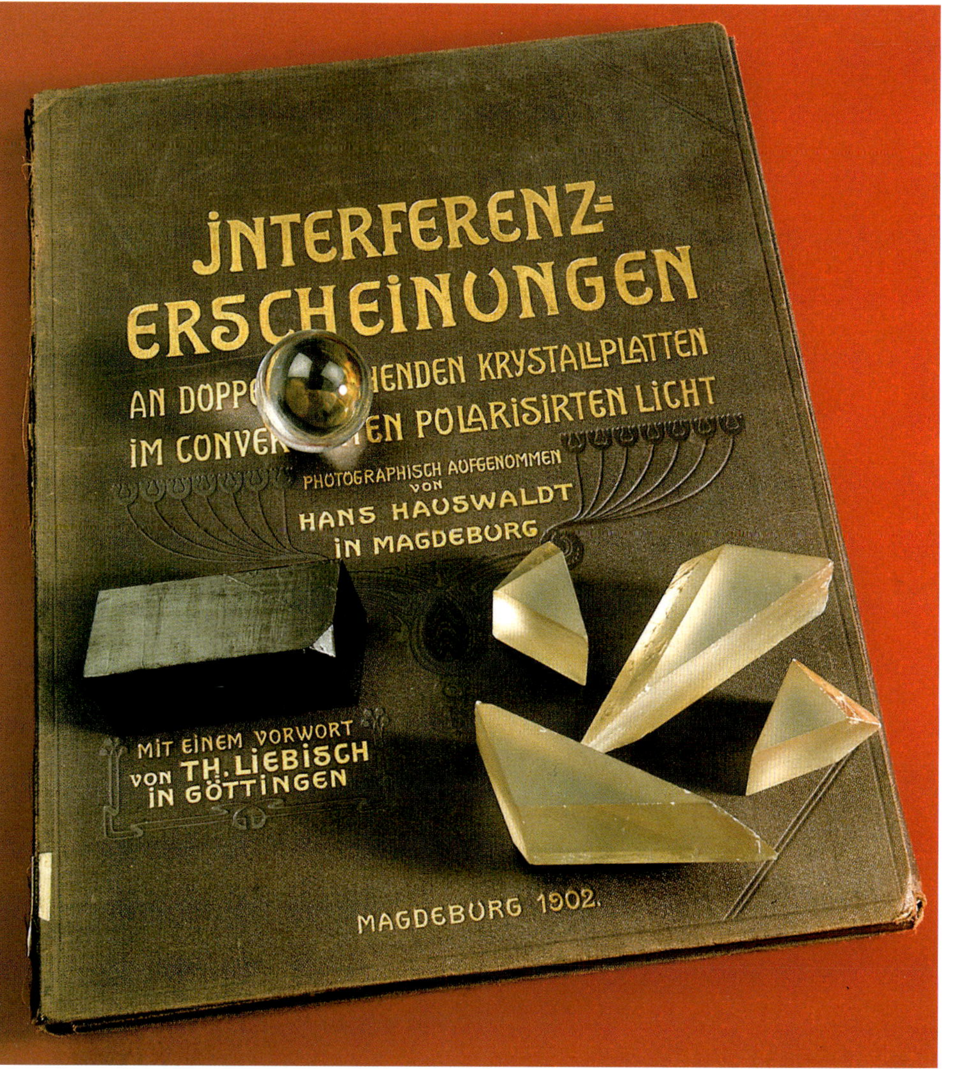

*Among the treasures found in the Natural Science Museum of Humboldt University in Berlin is Hans Hauswaldt's first edition (**right**) with equipment including a Glan-Thompson prism, the remainder of the crystal out of which the prism was cut and a calcite sphere. Photo Waltraud Harre*

*Candy manufacturer Johann Gottlieb Hauswaldt ("JGeHa") was founded by and named for Hans' father. The company used a variety of advertisements (**facing page**) to sell its sweets, these are now collectibles. Günter Grundman Archive*

Optics ◊ Finds ◊ Fluorescence

One Man's Contribution to the Advancement of Mineralogy

From the discovery of comets to mineral species, amateurs have traditionally played an important role in the advancement of science. Historically, scientists can ill afford to make the hundreds of precision measurements and analyses required to advance an idea. It is often thus left to inspired amateurs, gifted with both talent and time, to carefully observe and document subtle changes in the natural world.

In the late 1800s, reclusive German chocolate manufacturer Hans Hauswaldt of Magdeburg-Neustadt dedicated his off-hours energy to photographing interference figures, patterns that are created when light rays polarized by doubly refracting crystals (see page 40), such as calcite, are directed toward one another. The distinct patterns transmitted depend on the type of mineral, the direction of the incident light and the thickness of the sample. Petrographers look at thin sections of minerals through a modified polarizing microscope and compare the resulting interference pattern with a reference set to identify the minerals present in the thin section. Encouraged by mineralogist Theoder Liebisch, Hauswalt helped to establish this set of references.

Using monochromatic light, he created black and white rather than color patterns, which would not have been clearly reproduced on the film available at the time (color film had yet to be invented). To create the black and white patterns, Hauswalt designed an alcohol lamp and added sodium chloride to the fuel to generate yellow light. Combustion took place in pure oxygen, which needed to be controlled and adjusted with extreme care; still, he was able to achieve exposure times of up to sixty minutes. The apparatus conceivably set fire to his laboratory more than once.

Hauswaldt generated the polarized light with a 4 centimeter thick, Glan-Thompson calcite prism. To avoid distortion, he used a tourmaline plate as the analyst, instead of a second prism. He found that silver-eosine film plates produced by the Perutz company in Munich produced the best results. The thin-sections that he used were fashioned, with few exceptions, by Steeg and Reuter of Homburg.

Between 1897 and 1901, Hauswaldt prepared no less than 1,000 images. In 1902 he privately published the first series of thirty-three plates each with four images under the title *Interference Phenomena in Doubly Refracting Crystal Plates in Convergent Polarized Light, photographed by Hans Hauswaldt in Magdeburg*. Eighty plates followed in 1904; seventy-two additional plates made up a third series. Hauswaldt's images set a standard that has not been duplicated since. Rather than recreating patterns, many mineralogy and crystallography textbooks have adopted his images over the years and continue to use them today.

Condesenced from a full length article by ***Ferdinand Damaschun*** *that appeared in the German-language* extraLapis No. 14: Calcite

Two Polymorphs of $CaCO_3$:

Calcite

Composition
$CaCO_3$; for naturally occurring calcite, there is complete solid solution between Ca and Cd and limited substitution (typically on the order of a few percent) between Ca and Mg, Fe, Mn, Co and Ni. In the laboratory, complete solid solution has been produced between calcite and rhodochrosite.

Dana number: 14.1.1.1; Strunz: 5.AB.05

Crystal Structure
Calcite's structure consists of alternating layers of calcium atoms and carbonate groups stacked along the *c*-axis. Calcium has octahedral coordination with six oxygens from six different CO_3 groups.

Crystal System
Calcite is trigonal and is in the ditrigonal-scalenoedral crystal class, $\bar{3}\frac{2}{m}$. In the hexagonal setting, the lattice constants for the structural unit cell are $a = 4.99$ Å, $c = 17.06$ Å, $c{:}a = 3.419$. The traditional morphological unit cell has 1/4 the translation along *c* and *a* $c{:}a$ ratio of 0.8547.

Hardness
Calcite is the reference mineral for hardness 3 on the Mohs scale.

Cleavage
Perfect on $\{10\bar{1}1\}$

Fracture
Conchoidal, but difficult to produce

Luster
Glassy

Density
Pure calcite (*Iceland spar*) has a density of 2.71 g/cm^3. Other samples range in density from 2.6 to 2.8 g/cm^3, depending on impurities.

Streak
White

Color
Calcite is commonly colorless to white; other colors are relatively rare. When colored crystals are found, they are generally yellow to honey-colored. Substitution of Co for Ca results in the rose-red *cobaltian* variety; Mn substitution results in pink *manganoan* calcite. Green, red, blue, gray or black crystals are often the result of mineral inclusions.

Color Changes
Irradiation can create red-brown to black crystals.

Aragonite

Composition
$CaCO_3$; aragonite is generally very pure; however, small amounts of Sr, Ba, Pb, Fe, Mn and Zn sometimes substitute for calcium. The lead variety *tarnowitzite* contains up to 5 weight percent $PbCO_3$; this variety occurs as microscopic intergrowths with $CaCO_3$

Dana number: 14.1.3.1; Strunz: 5.AB.15

Crystal Structure
The structure of aragonite is similar to that of calcite. Layers of calcium ions and carbonate groups alternate along the *c*-axis. The structure shows pseudohexagonal symmetry; this near-hexagonal geometry facilitates the formation of twins and trillings. The structure differs from that of calcite primarily in the nine bonds between the calcium atoms and their neighboring oxygen atoms.

Crystal System
Aragonite is orthorhombic and is in the rhombic-dipyramidal crystal class, $\frac{2}{m}\frac{2}{m}\frac{2}{m}$. The lattice constants are $a = 4.96$ Å, $b = 7.96$ Å, $c = 5.73$ Å.

Hardness
3.5 to 4 on the Mohs scale

Cleavage
Distinct on $\{010\}$

Fracture
Uneven to conchoidal

Luster
Glassy, on fractured surfaces greasy

Density
2.94 g/cm^3; there are only slight deviations in the density of aragonite because there is very limited substitution.

Streak
White

Color
Aragonite is generally colorless to white. Other colors are rare. Colors in yellow, green, rose-red, violet, blue, gray and black crystals are typically caused by mineral inclusions.

Color Changes
Crystals can be irradiated red-brown to black.

Similarities and Differences

Calcite

Luminescence
Calcite fluoresces yellow, red, blue or orange under longwave as well as under shortwave UV. Some crystals exhibit multi-colored fluorescence, and some phosphoresce markedly after exposure to ultraviolet radiation

Refraction
Very high birefringence $n_o - n_e = 0.172$, $n_o = 1.6583$, $n_e = 1.4864$ for yellow Na light

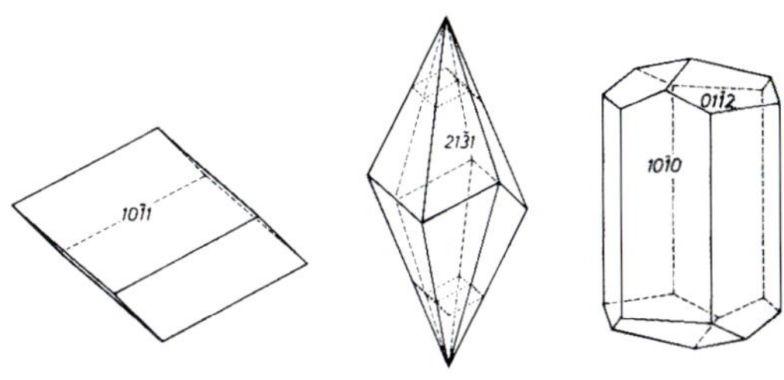

Characteristic crystals: rhombohedron (left), scalenohedron (middle), prism (right) (after Strunz, 1978)

Crystals
Crystals are common and form as various rhombohedra, scalenohedra and dipyramids as well as prisms and pinacoids and in numerous combinations. Twins form according to four twin laws and at some localities are common. Polysynthetic twins are caused by mechanical distortion.

Habit
Tabular, equant, prismatic, acicular, shallow to steep rhombohedral, uncommonly pseudocubic

Occurrence
Calcite is widespread occurring in igneous, metamorphic and sedimentary rocks and as crystals in cavities in nearly all rock types, including silicate rocks, pegmatites and hydrothermal veins. Calcite precipitates in hot springs, in caves and is even found in meteorites and in biologic settings.

Solubility
Calcite effervesces in all acids and dissolves readily in cold, dilute hydrochloric acid.

Similar Minerals
Perfect cleavage differentiates calcite from aragonite and strontianite both of which have poor cleavage. Dolomite and some other carbonates do not react to cold, dilute hydrochloric acid.

Aragonite

Luminescence
Fluorescence is common and crystals can be yellow, red, blue or orange under longwave UV. A single crystal will sometimes fluoresce in multiple colors. Some aragonite is markedly phosphorescent after exposure to ultraviolet radiation.

Refraction
Birefringence $n_z - n_x = 0.154$, $n_x=1.531$, $n_y=1.681$, $n_z=1.685$ for yellow Na light

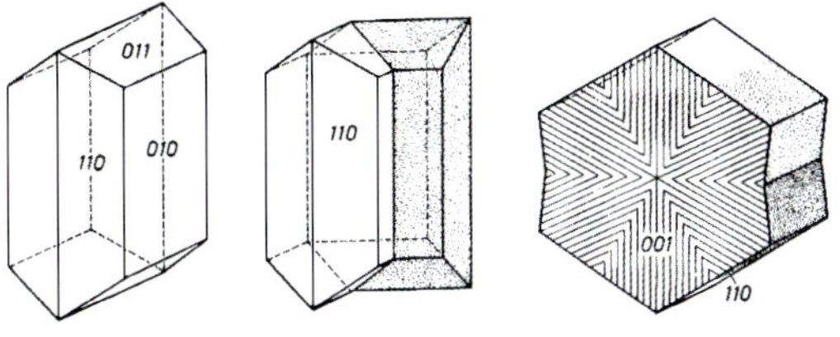

Characteristic crystal (left), twin (middle) and trilling (right)

Crystals
Crystals are common and form as various prisms, pinacoids and dipyramids. Some crystals have many faces, but finds of well-developed crystals are rare. Single crystals are uncommon. Twins and pseudohexagonal trillings, sometimes with polysynthetic twinning, are more common.

Habit
Crystals are very often elongated to acicular along the vertical axis, but may also be tabular or bladed.

Occurrence
Aragonite is found in cavities in volcanic rocks, as a high-temperature precipitate of seawater (e. g. Red Sea), in crystalline limestones of the prehnite-pumpellyite-facies, in sulfur deposits, as a precipitate in hot springs and in caves. The shells of many snails and clams are composed of aragonite.

Solubility
Aragonite effervesces in all acids and dissolves readily in cold, dilute hydrochloric acid.

Similar Minerals
Poor cleavage distinguishes aragonite from calcite and other rhombohedral carbonates. It is more readily soluble in cold, dilute hydrochloric acid than is strontianite and is much more soluble than other carbonates except calcite.

Calcite and Its Relatives

Museum curators ***Rupert Hochleitner****, Bavarian state collection in Munich, Germany and* ***Terry Huizing****, Cincinnati Museum Center in Cincinnati, Ohio with a look at the mineralogy of calcite and related species*

Through its history, *calcite*, the dominant member of the rhombohedral carbonates, has had numerous names such as *calcspar*, *Iceland spar* and others depending on habit, inclusions or environment of formation. The many beautiful and perfect crystals of calcite have fascinated scientists for hundreds of years and have initiated intense research into the composition and structure of this mineral.

Calcite and a number of other carbonates share the general formula $A(CO_3)$, in which *A* represents a divalent cation. These minerals crystallize with the structure of either calcite or aragonite. As illustrated in Table 1 below, those ions smaller than calcium produce minerals with a trigonal calcite structure, while those larger than calcium produce minerals with an orthorhombic aragonite structure; the size of the cation is, therefore, the critical factor in determining which structure is formed. Calcium is the only cation that occurs as a carbonate in both structures. Calcite and aragonite are thus known as *polymorphs*: they share the same chemical formula but have different crystal structures. A third naturally occurring polymorph is *vaterite*, calcium carbonate with a hexagonal structure. It is, however, a relatively uncommon mineral and of little interest to this review.

Table 1: *The relationship between cation size and structure for carbonates with the general formula* $A(CO_3)$

Element	Cation	Ionic Radius in Å	Calcite Group Trigonal Structure		Aragonite Group Orthorhombic Structure	
Magnesium	Mg^{2+}	0.66	Magnesite	$MgCO_3$		
Nickel	Ni^{2+}	0.69	Gaspéite	$NiCO_3$		
Cobalt	Co^{2+}	0.72	Sphaerocobaltite	$CoCO_3$		
Iron	Fe^{2+}	0.74	Siderite	$FeCO_3$		
Zinc	Zn^{2+}	0.74	Smithsonite	$ZnCO_3$		
Manganese	Mn^{2+}	0.80	Rhodochrosite	$MnCO_3$		
Cadmium	Cd^{2+}	0.97	Otavite	$CdCO_3$		
Calcium	Ca^{2+}	0.99	Calcite	$CaCO_3$	Aragonite	$CaCO_3$
Strontium	Sr^{2+}	1.12			Strontianite	$SrCO_3$
Lead	Pb^{2+}	1.20			Cerussite	$PbCO_3$
Barium	Ba^{2+}	1.34			Witherite	$BaCO_3$

At least some substitution is known to occur between members of the calcite group. For naturally occurring calcite, there is complete solid solution between calcium and cadmium, and only limited solid solution (typically on the order of a few percent) between calcium and magnesium, iron, manganese, cobalt and nickel (Reeder, 1983). These substitutions produce some of the more colorful calcite varieties, such as rose-red *cobaltian* calcite and pink *manganoan* calcite. Even large cations, such as barium, lead and strontium that ordinarily enter into aragonite-type structures, substitute in small amounts for calcium in calcite and produce varieties such as *plumbian* calcite. Pure calcite is either colorless or white, but given the variety of ions that can enter into its structure, it is not surprising that calcite exhibits a range of colors and habits.

In the aragonite group there is limited solid solution between members. For aragonite itself, only a small amount of strontium or lead and no barium is reported to substitute for calcium (Gaines et al., 1997). Even supposed solid solutions such as *tarnowitzite*, the lead variety of aragonite, have been shown instead to be microscopic intergrowths (Strunz, 1967).

Other well known carbonates such as dolomite, ankerite and kutnohorite are double carbonates whose structures are derived from symmetry reductions of the calcite structure. These minerals are included in the dolomite group and are not included in this summary.

Thousands of Habits

More than 600 forms have been documented for calcite (Palache, 1943). As a single crystal of calcite may be bounded by more than one form, it is not uncommon for three, four, or even five or more forms to be combined on a single crystal; thus, the number of documented form combina-

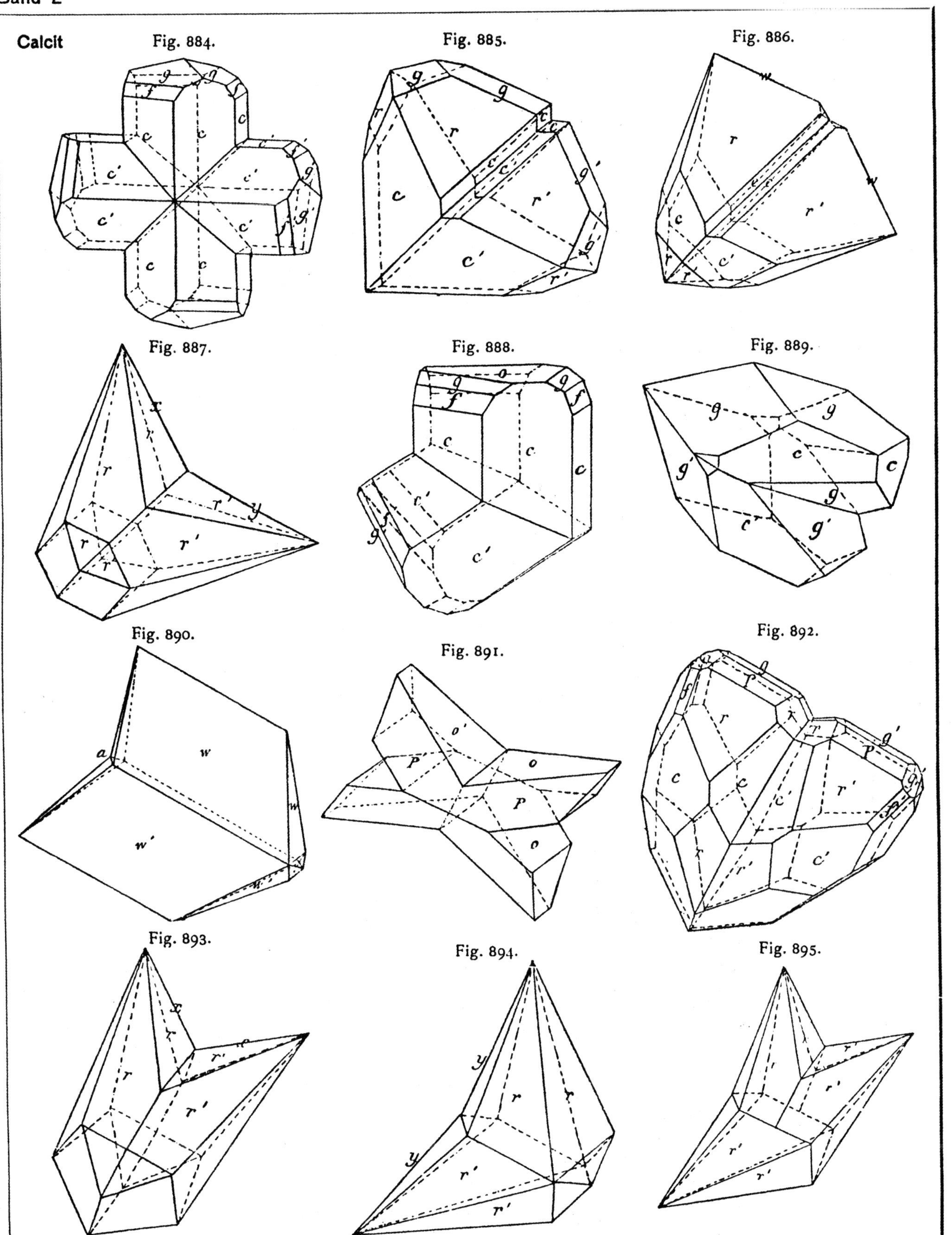

Localities:

Fig. 884:
Bräunsdorf, Saxony, Germany

Fig. 885, Fig. 886, Fig. 887:
Derbyshire, England

Fig. 888:
Harz Mtns., Lower Saxony, Germany

Fig. 889:
Freiberg, Saxony, Germany

Fig. 890:
Faroe Islands, Denmark

Fig. 891:
Chamonix, France

Fig. 892, Fig. 893, Fig. 894, Fig. 895:
Bleiberg, Austria

One of no less than 154 plates depicting a total of some 2700 calcite habits that appear in Victor Goldschmidt's* Atlas der Krystallformen *(Heidelberg, 1913).

This plate includes some of the twins sketched by Austrian mineralogist Wilhelm Karl von Haidinger (originally published in 1825 in the Edinbourgh Journal of Science*)*

The Mineral with the Most Forms

Rhombohedron
Calcite is known to take shape as more than eighty different rhombohedra, varying from obtuse to acute.
This 6.1 cm wide specimen from Tsumeb, Namibia is in the collection of Francis Benjamin.
Photo Jeff Scovil

tions seen in calcite reaches into the thousands. While this variety of combinations is greater than that of any other mineral, all of the forms of calcite fall into five groups. Two groups contain open forms: the prisms and the pinacoid. Three contain closed forms: the rhombohedra, scalenohedra and dipyramids. Some of these groups of forms are highly variable. For example, more than 80 different rhombohedra, ranging from very flat (obtuse) to very steep (acute) and more than 200 different scalenohedra have been described.

Five Basic Forms

Pinacoid - An open form made up of two parallel faces that are both perpendicular to the *c*-axis

Prism - An open form composed of six or twelve faces, all of which are parallel to the *c*-axis

Rhombohedron - A closed form composed of six faces; three faces at the top of the crystal alternate with three faces at the bottom, the two sets of faces are offset by 60 degrees

Scalenohedron - A closed form with twelve faces grouped in symmetrical pairs, three pairs above and three below in alternating positions; in perfectly developed crystals, each face is a scalene triangle; the faces meet in a zigzag line around the girdle of the crystal

Dipyramid - A closed form having twelve faces, six on the top of the crystal and six immediately below them on the bottom; each face is an isosceles triangle

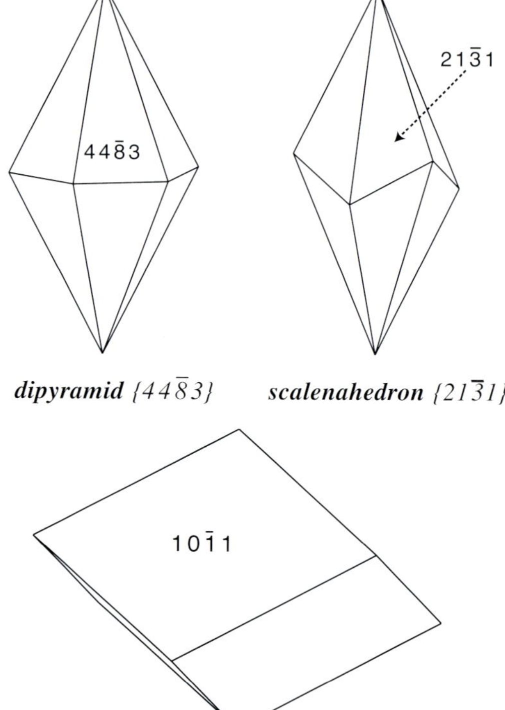

dipyramid *$\{44\bar{8}3\}$* ***scalenahedron*** *$\{21\bar{3}1\}$*

the unit ***rhombohedron*** *$\{10\bar{1}1\}$*

Hundreds of variations and combinations of the five basic forms give rise to the thousands of habits that are seen in calcite crystals

Drawings by Pete Richards using SHAPE software

There are thus a tremendous variety of calcite *habits*, defined as the characteristic form or combination of forms that result in the general shape of the crystal. There are several commonly observed habits for calcite: prismatic crystals with long or short prisms in which the prism faces are prominent and with either a pinacoid or rhombohedral termination; rhombohedra in which oblique to obtuse rhombohedral forms predominate; scalenohedra, often with prism faces and rhombohedral terminations; and thin-to-thick tabular crystals in which the pinacoid is prominent. Habits of common crystal aggregates may be described as fibrous, earthy (chalky), nodular (oolitic) and stalactitic. Clusters that reflect oriented (epitactic) overgrowths of secondary calcite are common. The variety of calcite habits is even more remarkable considering that the other members of the calcite group show only ten or fifteen different habits (Kostov, 1968). The great number of habits is the likely result of the array of conditions under which calcite forms.

For calcite, the development of crystal faces seems to be related, at least in part, to the nature of the parent solution, which varies based on numerous factors such as supersaturation (due to changing temperature), the proportion of calcium

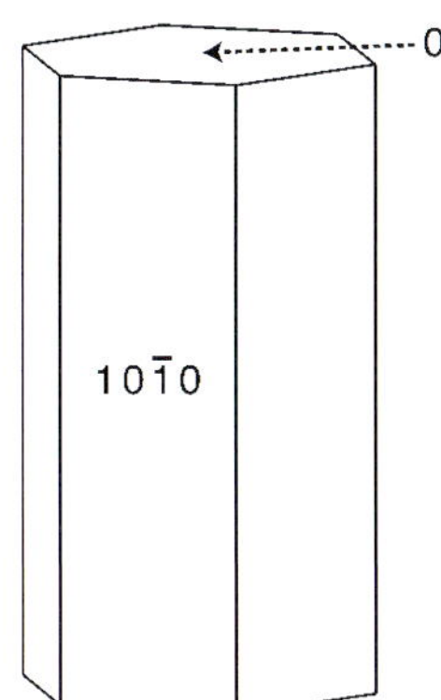

pinacoid *{0001} and* ***prism*** *{10$\bar{1}$0}*

As the pinacoid and prism are both open forms, they always occur in combination with one another or with other forms.

to carbonate ions, impurities and pH or salinity (Kostov, 1968). For example, elongated steep rhombohedral crystals occur when cool solutions have more calcium than carbonate ions, while tabular crystals with prominent pinacoid faces occur when hot solutions have excess carbonate ions. Interestingly, the unit rhombohedron {10$\bar{1}$1}, the most common form for all the other rhombohedral carbonates, is rarely found as the dominant form in calcite. Growth experiments suggest that the rhombohedron develops only when calcium and carbonate ions are present in nearly equal concentrations (Reeder, 1983). Many of the conditions that influence the variety of calcite habits are poorly understood, and in most cases, it is not yet possible to explain why a complex calcite crystal formed.

Twinning

When two or more crystals grow in fixed rather than random relationship to one another, they are described as *twins*. Because these relationships are related to the atomic structure of each mineral, the number of laws under which twins can form ranges from none to several. Calcite twins are known to form under four *twin laws*; twice as many twin configurations as for any of its relatives. Typically, twins are uncommon and are highly prized by collectors.

Twins are unknown or uncertain for five members of the calcite group including gaspéite, magnesite, otavite, smithsonite and sphaerocobaltite. In contrast, within the aragonite group, cerussite and witherite almost universally crystallize as twins, and aragonite often occurs as twins. The described twin laws for calcite and related minerals are summarized below.

In Classic Form

Above: *This group of calcite* ***scalenohedra*** *(the largest crystal is 4.3 cm long) is from the Droujba mine in Laki, a small town set in Bulgaria's Rhodope Mountains. Keith Williams collection; Jeff Scovil photo*

Left: *In the late 1970s, Neal Pfaff collected this 4 cm long calcite* ***dipyramid*** *from the Pugh quarry near the town of Custar in Wood County, Ohio. Terry Huizing photo*

Bottom left: *This 7.7 cm high hexagonal crystal from Charcus in San Luis Potosi, Mexico combines beautifully the open* ***pinacoid*** *and* ***prism*** *forms. Steve Smale collection and photo*

Eight different crystal forms are combined in this calcite habit:

4 different rhombohedra,
2 different scalenohedra
and 2 different prisms.

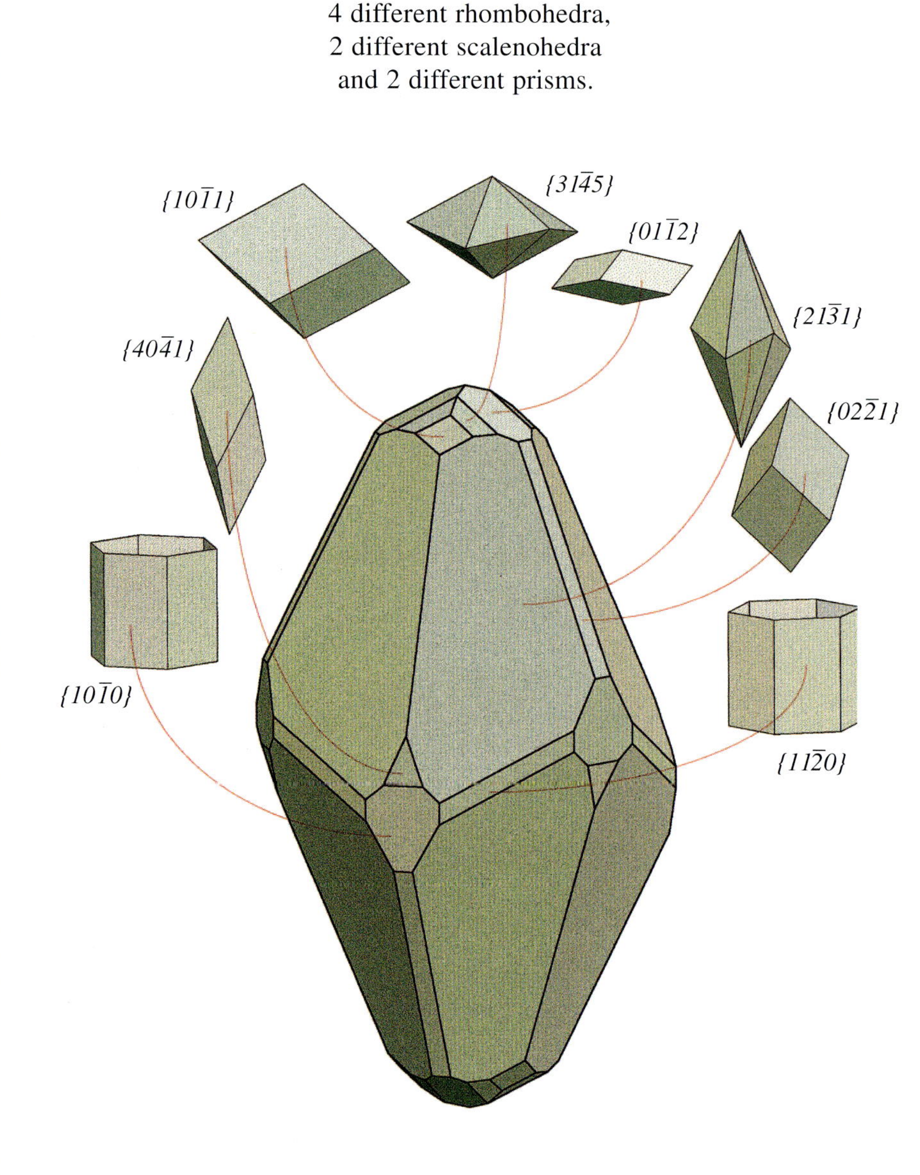

The illustration was drawn by Erich Offerman of Arlesheim, Switzerland using SHAPE software

Aragonite - Practically all twinned crystals are repeatedly twinned along a plane parallel to the prism {110} resulting in pristmatic crystals that are nearly hexagonal on cross section. The twinned nature of these crystals may often be seen as reentrant angles (notches) between adjacent prism faces and by striations on the basal pinacoid. Twinning has been doubtfully observed on {103} (Gaines, 1997). Exceptional pseudohexagonal twinned prisms of aragonite occur at the sulfur mines in Sicily, Italy, at Molina de Aragon, Spain, at Frizington, Cleator Moor and other places in Cumbria, England.

Calcite - There are four twin laws for calcite (see Table 2). For one the *c*-axes of the two twin components are at 180° to one another; under the other three laws, the twin axes are inclined at an angle to one another. Inclined-axis twinned crystals are often described by inexact terms such as *butterfly*, *heart-shaped* or *axe-head* twin. These terms describe the appearance of the twin, but not the twin law. Calcite twins are thus best referred to by the Miller indices of the shared plane across which the twinned parts are joined as mirror images of one another. The habit of twinned calcite is commonly quite different from the habit of untwinned crystals (Richards, 1999). Rapid growth occurs on faces where the two parts of the twin meet resulting in modifications of the crystal shape.

• Twins on the basal pinacoid **{0001}** are the only ones for which the twin plane is perpendicular to the *c*-axes of both parts of the twin and for which the *c*-axes are at 180° to one another (Richards, 1999). These twins are easy to recognize by reentrant notches along the contact

Morphologic vs. Structural Unit Cell

The Miller indices $\{10\bar{1}1\}$, used to describe the plane that is parallel to the cleavage of calcite, are based on the morphologic unit cell. This was defined before the use of X-ray diffraction determined the actual unit cell of the calcite structure. The actual Miller indices for this plane have since been determined to be $\{10\bar{1}4\}$ (Reeder, 1983). Morphological indices are used throughout this article for historical continuity. To convert Miller indices for calcite from morphologic to structural units, simply multiply the l index of {hkil} by four.

Table 2: The Four Twin Laws of Calcite
(in order of decreasing angle between the *c*-axes)

Morphological Unit Cell	Structural Unit Cell	Angle Between the *c*-axes
{0001}	{0001}	180°
$\{01\bar{1}2\}$	$\{01\bar{1}8\}$	127°30’
$\{10\bar{1}1\}$	$\{10\bar{1}4\}$	90°46’
$\{02\bar{2}1\}$	$\{01\bar{1}2\}$	53°46’

at the twin plane. This is the most common twin law for calcite; specimens of this type come from numerous localities. Recently, some of the most dramatic examples have come from the Elmwood mine near Carthage in Smith County, Tennessee, a locality that has produced world-class specimens (see page 76).

• If the *c*-axes of the twin are inclined at an angle of 127°30' with respect to one another, the twin plane is parallel to a face of the shallow negative rhombohedron **{01$\bar{1}$2}**. This is the second most common twin law for calcite, and representations are found at many localities. Some of the most notable include the mines near Egremont in Cumbria, England; the Verchniy mine and others around Dal'negorsk in Russia's Primorskiy Kray; as well as Porvenir and other mines from Areponapuchic, Chihuahua, Mexico. This type of twinning is also produced by mechanical deformation, as when calcite is metamorphosed to marble. The typical striations observed on such calcite evidence repeated twinning.

• When the *c*-axes of the twin are inclined at an angle of 90° 46' with respect to one another, the twin plane is parallel to a face of the positive rhombohedron **{10$\bar{1}$1}**, which is defined by the cleavage of calcite. This calcite twin law is rare compared to the two described above. Notable localities for this type of twin are few but include the Leiping mine in Guiyang County in the Hunan Province of China (see page 48); the hematite mines of West Cumbria, England, where this twin law is common; and Wheal Wrey, Liskeard in Cornwall, England (see page 64).

• If the *c*-axes of the twin are inclined at an angle of 53°46' with respect to one another, the twin plane is parallel to a face of the steep negative rhombohedron **{02$\bar{2}$1}**. This is another rare twin law for calcite. Several localities have produced notable specimens in Mississippi Valley-type (MVT) deposits of the midwestern United States. These include the Tri State mine at Cardin, Ottawa County, Oklahoma in the now-abandoned Tri-State lead-zinc district and the Brushy Creek mine in the Viburnum Trend district, Reynolds County, Missouri.

Cerussite $PbCO_3$ – Crystals are commonly found in pseudohexagonal groups or reticulated aggregates due to twinning on planes parallel to the prisms {110} and less commonly as *heart* or *v-shaped* contact twins on {130}. Magnificent *v-shaped* twins were found at the Touissit mine in Morocco, and fine reticulated twins of exceptional quality occurred at the Tsumeb mine in Namibia

***Above:** Hunan, China has become famous for producing calcite twins. This 9 cm wide **{10$\bar{1}$1} twin** is from the Leiping mine in Guiyang County Hunan.*

Steve Neely collection; Jeff Scovil photo

***Above:** The reentrant notches at the contact with the twin plane make this 8.2 cm wide Elmwood, Tennessee specimen easily recognizable as a **{0001} twin**. Collection Bill Severance; photo Jeff Scovil*

***Below:** This 6 cm wide **{01$\bar{1}$2} twin** from the Wessels mine, Cape Province, South Africa, is in Terry Huizing's collection who is also credited with the photograph.*

In Rare Form

Calcite rarely forms ***$\{0\bar{2}21\}$*** *twins, though in 1994, the Brushy Creek mine in Reynolds County, Missouri produced a find of some 40 specimens. This fine 5.5 cm tall example is owned and was photographed by Terry Huizing.*

and the Proprietary mine at Broken Hill in New South Wales, Australia.

Rhodochrosite $MnCO_3$ – Twins are rare, and twinning is parallel to the negative rhombohedron $\{01\bar{1}2\}$. Large color-zoned rhombohedra, twinned on the basal plane {0001}, were found in Canada at Mont Saint-Hilaire, Québec.

Siderite $FeCO_3$ – Twinning is uncommon parallel to the negative rhombohedron $\{01\bar{1}2\}$ and rare on a plane parallel to the basal pinacoid {0001}. Large tan rhombohedra modified by the basal pinacoid and twinned on {0001} have been locally common at Mont Saint-Hilaire.

Strontianite $SrCO_3$ – Twinning is common parallel to the prism {110}. Crystals usually occur as contact twins, rarely as penetration twins. Repeated twinning, particularly in high calcian varieties, results in pseudohexagonal aggregates that resemble those found in aragonite.

Witherite $BaCO_3$ – Twins are almost universal and are most common parallel to the prism {110}, producing sharp pseudohexagonal dipyramids. Typically, many such dipyramids grow together as stubby pseudohexagonal prisms with stepped sides and curved tops. The finest of these have been found in the United States at the Minerva No. 1 mine in Hardin County, Illinois and previously from northern England, particularly at Alston Moor in Cumbria.

Calcite – Found Worldwide

Calcite is a widely occurring mineral; it is very common and abundant in all classes of rocks except granitic types, which are poor in calcium. Calcite is found in **igneous environments**, in cavities and fissures in basalts and related rocks. Notable examples are the attractive crystals that occur, often with other minerals, in frozen gas bubbles and other voids in flood basalts in the Brazilian state of Rio Grande do Sul (see page 84) and throughout the Deccan traps of Maharashtra, India (see page 57). At Helgustader, Eskifjord, Iceland, calcite occurred in cavernous basalts as an enormous deposit of optical calcite, including some crystals to 7 meters (!) in length. Beautiful crystals, sometimes included with brilliant native copper, occur in the basalt flows of the Keweenaw Peninsula of Michigan. In all cases, these crystals formed after the solidification of the host rock.

Unusual occurrences of calcite include *carbonatites*, ultrabasic alkaline magmatic rocks that can be almost entirely composed of calcite. The deposit at Kaiserstuhl in Germany contains crystals of calcite, while in the East African Rift, liquid carbonatite magma from the Oldoinyo volcano has flowed over the surface; it is a fascinating, almost unreal sight to see white lava!

Magnificent calcite crystals are found in polymetallic deposits at Dal'negorsk in Primorskiy Kray, Russia (see page 60), where skarn bodies of hedenbergite composition are a product of limestone replacement. Here, in a karstlike environment, a variety of beautiful calcite crystals occur and are found in a number of habits including twins, as the pink *manganoan* variety and in association with other wonderfully crystallized minerals.

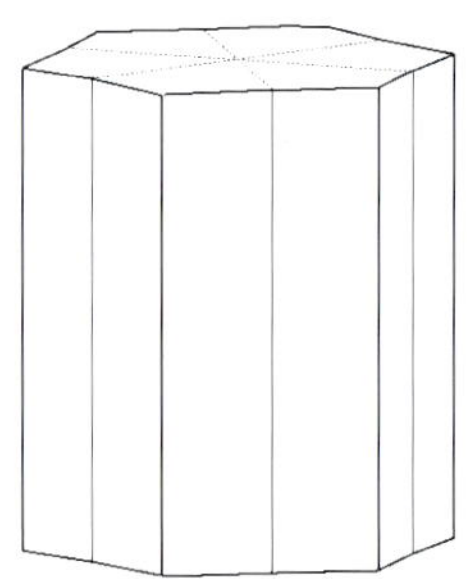

Aragonite penetration trilling on {110} Aragon, Spain

Drawings by Pete Richards using SHAPE software

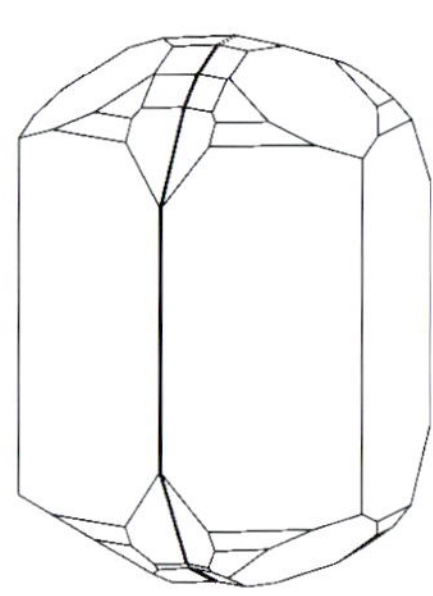

Strontianite twin on {110} Clausthal, Germany (after Dana VII)

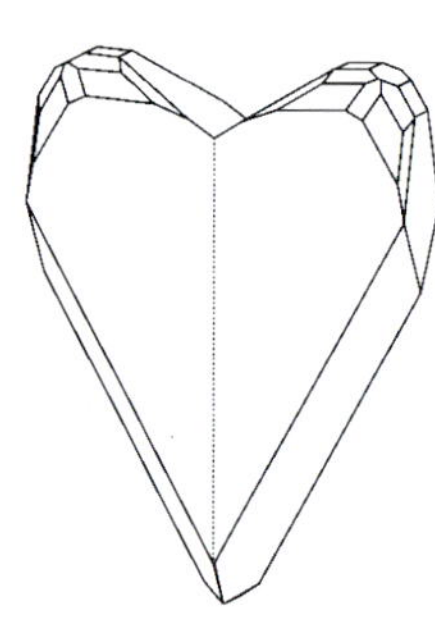

Cerussite twin on {130} Tsumeb, Namibia (after Dana VII)

1

Calcite Is Not Always Colorless!

1. Ray mine, Pinal County, Arizona

The color of this 5.4 cm high specimen is the result of cuprite inclusions. McCarty collection

2. Tsumeb mine, Tsumeb, Namibia

The calcite in this specimen owes its color to inclusions of malachite. Judy Megerle collection

3. Herja mine, Maramures, Baia Mare, Romania

The color of the top half of this 5.6 cm ball has been variously attributed to jamesonite and to boulangerite. Victor Yount collection

4. Dikulukwe mine, Kolwezi, Shaba Province, Zaire

The element cobalt colors this 6 cm high crystal of cobaltian calcite. Roberts Minerals

5 Sokolovskiy-Sarbaiskiy mine, Rudniy, Khazakstan

A 7.9 cm calcite twin on {0001} whose yellow color is caused by iron. Bill Shelton collection

All photos Jeff Scovil

2

3

4

5

Calcite is often a gangue mineral of hydrothermal veins, and fine crystals may be found in openings within orebodies exposed during commercial mining for metals. Notable localities include the Tsumeb mine in Namibia, where sharp but simple rhombohedra occur, and at Saint Andreasberg in the Harz Mountains of Germany (see page 72). At the San Sebastian mine at Charcas in San Luis Potosi, Mexico, large tabular crystals with prominent pinacoid faces occur. In the mining districts of Mapimi, Durango as well as in Santa Eulalia, in Areponapuchic and in Chihuahua, Mexico, splendid crystals occur in limestone replacement deposits within cavelike structures in manto and chimney orebodies.

In **sedimentary environments**, beautiful formations of calcite commonly occur in limestones as long as some type of void is provided for crystallization. These formations may range from speleothems, calcite (or aragonite) deposited in caves by the action of meteoric water, (see page 94) to crystals deposited in vugs and other open spaces by the action of other fluids passing through the host rock.

When these fluids deposit economically significant elements, as in lead-zinc Mississippi Valley-type (MVT) deposits or in fluorspar deposits, the workings often expose associated calcite crystals, which tend to occur in a variety of habits, a range of twin laws and in attractive association with other minerals. Some MVT localities notable for fine calcite include the Elmwood mine in Smith County, Tennessee and the Viburnum Trend mines in Missouri. The fluorspar mines in southern Illinois and northern England have also produced world-class calcite specimens.

Notable crystals of calcite may be found when rock quarries expose voids within ancient reef structures such as those that occur in the limestones of the midwestern United States. Here the dipyramid form, reported as rare for calcite at most other locations, is found as a relatively common habit.

Calcite also occurs in **metamorphic environments** as marble. Marble is metamorphosed limestone; it is a granular rock composed of crystalline calcite grains. The most desirable marble is white, massive and shows no trace of cleavage due to its fine grain size.

Practical Uses of $CaCO_3$

The word *calcite* has its origins in Greece, where limestone was known as *chalix*; this word was adopted as *calx* in Latin. In 1808, Sir Humphrey Davy isolated a new element from *quicklime* by means of an electrolytic process and named it *calcium.* Then in 1845, mineralogist Wilhelm Karl von Haidinger applied the name *calcite* to all occurrences of calcium carbonate.

The close relationship between limestone, calcium carbonate and calcite is evidenced by the loose manner with which the terminology is applied. As described above, *limestone* is a sedimentary rock type composed mainly of calcium carbonate; *calcite*, *aragonite* and *vaterite* are the crystalline forms of calcium carbonate. *Chalk* is a friable form of limestone, and *lime* is a generic term that refers to *limestone* ($CaCO_3$), *quicklime* (CaO) and *hydrated lime* (CaOH). All three are different derivations of calcium carbonate; all having been known and used since ancient times.

Twenty percent of the Earth's sedimentary rocks are limestones, which form the basis for numerous inorganic chemical products (Schröcke-Weiner, 1981). Calcium carbonate, in the form of limestone or chalk, is used for the production of Portland cement and, sold as *lime*, is used to adjust the pH of soil. Heating or *calcining* calcium carbonate to high temperatures produces calcium oxide (*quicklime*, *calcined* or *burnt lime*). It has a number of practical uses that include buffering the pH in sewage treatment by increasing the alkalinity of the raw sewage and slowing the reaction during the manufacture of concrete to keep temperatures low. Calcium oxide is also used to desulfurize crude iron. Hydrating calcium oxide produces calcium hydroxide (*slaked* or *hydrated lime*), a compound used to provide elasticity in mortars, neutralize acidic pH levels, soften water and precipitate soluble and gaseous impurities, including acids and heavy metals, from certain solutions.

Glass production involves the formation of calcium silicate, made by melting quartz, sodium carbonate and calcium carbonate. Heating calcined lime with coke produces calcium carbide, an intermediate in the production of acetylene and the fertilizer calcium cyanamide. Calcium carbonate is added in the smelting of acidic iron ores in blast furnaces to obtain a low viscosity slag. Calcium compounds are generally used for deoxidizing molten metal and for the calcio-thermic reduction of numerous metal oxides to produce such metals as titanium, beryllium, zirconium, molybdenum, tungsten, thorium and uranium.

Since ancient times, civilizations have used limestone as a construction material: the pyramids of

χαλιξ · calx · Kalkspat · Lime · Calcite

Namibian Rarity
A stunning specimen of calcite (3 cm) on native copper. The calcite is included with cuprite and is from the Onganja mine, near Seeis. Carolyn Manchester collection; photo Jeff Scovil

Giza are all built from local limestone. Marble has been used for works of art and official buildings for millennia. The Lincoln Memorial in Washington, D.C. is constructed of a beautiful white marble from the Yule quarry at Marble in Gunnison County, Colorado. The fine-grained and thinly layered limestone of the Solnhofen area in Bavaria, Germany advanced the development of lithography.

The Role of Calcite

Well-crystallized calcite has repeatedly played a role in the history of crystallography and physics as a material for experimental investigations and as the subject of theoretic ones. Erasmus Bartholinus, for example, observed double refraction in *Iceland spar* in 1669. Bartholinus' discovery was the basis for Christian Huygens' 1678 explanation of the propagation of light based on wave theory. It also laid the foundation for crystal optics, especially polarized light microscopy. Calcite was used for making optical polarizers and Nicol prisms (see page 40). These were once expensive and very much in demand, but today they have been replaced by less expensive and more easily produced Polaroid sheets.

Calcite cleavages formed the basis for René Just Haüy's theory that crystals formed from elementary spatial units, a concept that inspired modern structural crystallography (see page 26).

While calcite may have lost its technical significance, it remains the darling of the specimen collector. Calcite crystals are objects of beauty and complexity and can be collected in many places throughout the world.

Crystal Shapes: A Primer

Crystallographer and calcite aficionado ***Pete Richards*** *walks us through the fundamentals of crystallography. Pete developed the Macintosh version of the SHAPE sofware, and all of the drawings illustrating this article were drawn by him using this software.*

A crystal is a solid with a homogeneous chemical composition, and unless constrained during growth by its environment, is bounded by naturally formed plane faces. It is composed of one or more kinds of atoms that are bonded to one another and arranged in a geometric pattern that extends and is repeated in all directions throughout the crystal. Everywhere in the crystal, the same arrangement of atoms is found and in the same orientation. The plane faces of the crystal reflect and are the result of this regular arrangement of the atoms that make up the crystal.

Contemporary crystallographers use X-ray techniques to identify minerals and determine their crystal structure. Before the early 20th century development of X-ray crystallography, however, studies of the external morphology of crystals rather than properties disclosed by X-ray analysis, were critical for describing new species, for identifying new examples of known species and for understanding the internal structure of crystals.

Over several centuries and long before the nature of atoms and most aspects of chemistry were understood, morphological crystallographers discovered fundamental properties of crystals and developed techniques for accurately measuring and describing their geometry. Morphological properties are still important aspects of mineralogy and crystallography.

The Unit Cell

Each fragment of a given crystal contains the same regular arrangement of atoms. Only a small fragment of the crystal is required to determine this arrangement - this template for building the crystal. Imagine, as did René Just Haüy (see page 26), dividing a crystal into smaller and smaller pieces until removing one more atom would render the template incomplete. This smallest unit of a crystal is known as the *unit cell*, Haüy's *molécule intégrante*.

A crystal can be thought of as being built up of very large numbers of unit cells, stacked in three dimensions. The unit cell defines the natural geometry of the crystal, including the orientation of symmetry axes relative to one another and the length of the unit step along each axis. The length of the unit step is often referred to as the length of the axis itself. The natural geometry of crystals differs from the Cartesian coordinate system studied in high school: in crystals, the three axes are not necessarily at right angles to one another and the unit steps along each axis are not necessarily the same size. The shape of the unit cell determines these axes and angles.

Not all crystals are minerals, but all minerals are crystals because all minerals are composed of specific atoms systematically arranged in a specific way. This atomic configuration - the crystal structure - is responsible for the combination of properties that make a mineral species unique: chemical composition, density, hardness, luster, cleavage, optical properties and often even color, though color can be strongly influenced by impurities and imperfections that are not essential to the crystal structure.

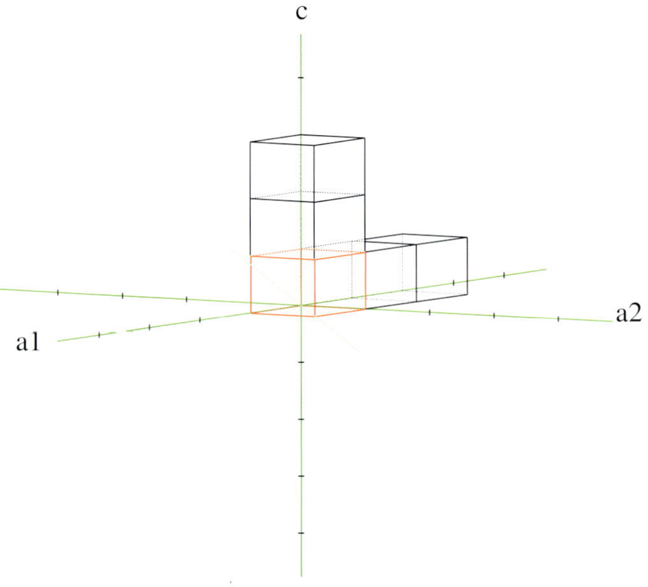

A unit cell (shown in red) and the natural coordinate system for calcite. Additional unit cells are stacked on top of the unit cell along the c-axis with two more along the negative direction of the first a-axis (a1). The crystal axes are extensions of the unit cell edges. Each mark along the axes represents the length of one unit cell step. The a-axes are perpendicular to the c-axis, and at angles of 120 degrees to one another. The third a-axis is represented by a faint line; it is not needed to describe the symmetry of calcite.

Symmetry

All crystals are characterized by *symmetry*: a set of geometric relationships that are expressed in the external form of the crystal, in the arrangement of its atoms, and in its optical and physical properties. All crystals of a given mineral have the same symmetry. Mirror planes, rotation axes and the inversion center are among the elements of symmetry that are used to describe the external form of crystals. Faces of a crystal that are related by a symmetry element are said to be the *symmetric equivalents* of one another.

In nature, crystals are distorted by uneven growth in different symmetrically equivalent directions. While the crystal structure may be symmetric, the growth environment is not! The most substantial distortion for most crystals results from attachment to the matrix, which generally allows for the expression of only about half of the crystal.

Crystallography describes ideal crystals that are free of such distortions. *Floater* crystals that have grown in a soft, uniform matrix such as mud or a gel tend to be nature's closest approximation to the crystallographic ideal. These ideals are implicit in the following descriptions.

A *mirror plane* is an imaginary planar surface inside a crystal that has the geometric properties of an ordinary mirror - every part of the crystal on one side of the mirror is matched by a corresponding part on the other side, an equal distance away measured along a direction perpendicular to the mirror. In contrast to images in the mirrors of our common experience, the parts on both sides of the mirror are equally real. Also in contrast to common experience, crystallographic mirror planes are most apparent (at least in drawings) when the two parts of the crystal related by the mirror plane are shown as if one were looking along the edge of the mirror, rather than directly into it. A mirror plane is often shown as a line, representing a plane oriented perpendicular to the plane of the drawing.

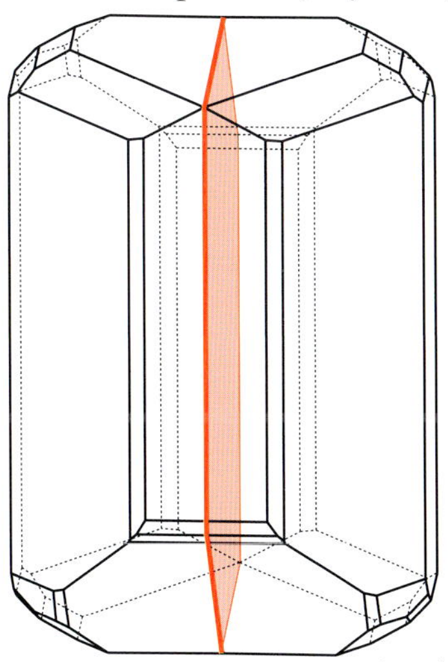

An ideal colemanite crystal with prominent mirror plane symmetry; the crystal's single mirror plane is shown in red

A *rotation axis* is a line about which a crystal can be rotated through a certain angle, such that the crystal looks exactly as it did before rotation. For example, a calcite crystal rotated by 120° or 240° about the *c*-axis looks the same at each stop as it did before starting the rotation.

The axes defined by the edges of the unit cell can always be considered to be rotation axes, though for some minerals the only rotation about a given axis that leaves the crystal looking unchanged is the trivial rotation through 360°, a complete circle. Some minerals also have rotation axes that are not defined by the edges of the unit cell.

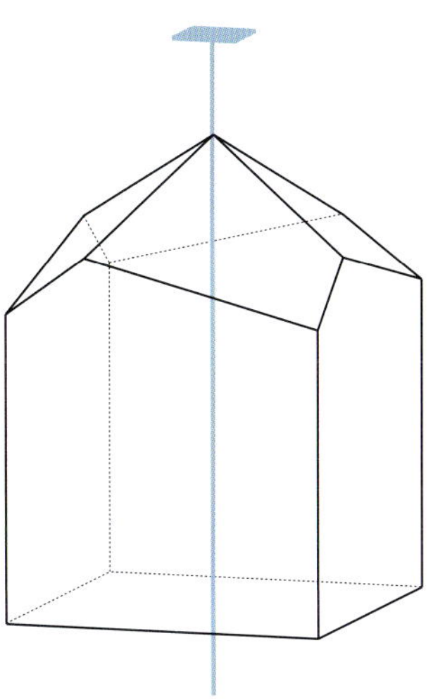

***Above**: A hypothetical crystal with four inclined faces at the top, and four vertical faces on the sides. Each face of either group is equivalent to all the other faces of the same group. The vertical axis is a 4-fold axis of rotational symmetry.*

Returning to the calcite example, there are three positions of rotation that leave the crystal looking the same: 120, 240 and 360 degrees. This property is described by indicating that calcite's *c*-axis is a *3-fold axis of rotational symmetry*. If a crystal has n-fold rotational symmetry, it looks the same at every rotation through 360/n degrees.

Crystals can only have 2, 3, 4 or 6-fold rotational symmetry. Five-fold symmetry and all symmetries 7-fold and higher cannot occur in crystals, as they would require unit cells that could not be stacked in three dimensions.

The *inversion center* is a point at the center of a crystal. If a mineral has an inversion center, its crystals have the characteristic that for any point on one side of the crystal, there is an identical point on the opposite side of the crystal that can be found by drawing a line from the first point through the inversion center to the other side of the crystal. This property applies to any point on the surface of the crystal, including all crystal corners, all points along the crystal edges and any point on any crystal face. Calcite is a mineral the crystals of which have an inversion center.

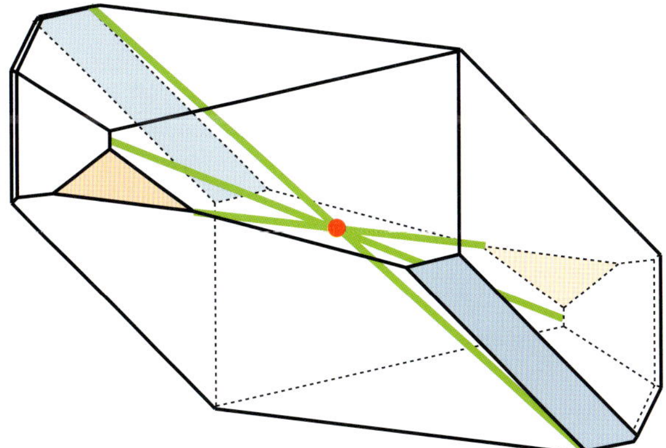

An albite crystal with its inversion center indicated by a red dot. Every point on the crystal is identical to a point on the opposite side of the inversion center. A few examples are indicated by green lines that connect equivalent points. Two colored faces correspond to similarly colored faces on the other side of the crystal.

Below: A calcite crystal showing the 3-fold rotation axis (vertical) and three 2-fold rotation axes (elliptical ends): the minus on one of the axes indicates its negative end. One of the mirror planes is shown and is indicated by the blue line.

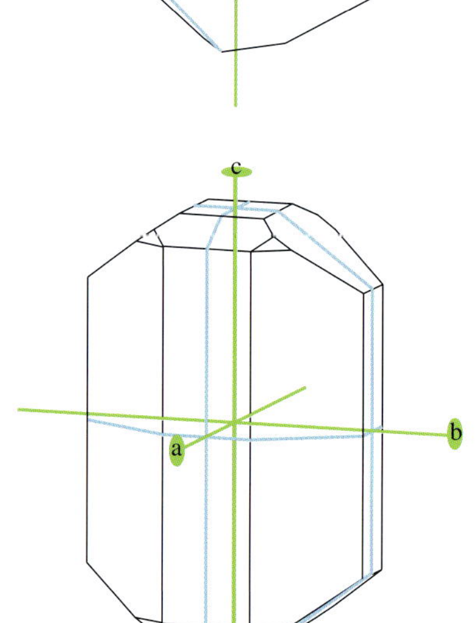

Above: *An aragonite crystal showing the positions of the a, b and c-axes. All three meet at right angles and all are 2-fold rotation axes. Each axis has a mirror plane (blue lines) that is perpendicular to it.*

Symmetry Systems and Classes

There are six systems of symmetry: triclinic, monoclinic, orthorhombic, tetragonal, hexagonal and isometric. The hexagonal system is subdivided into the hexagonal and trigonal subsystems. The six systems of symmetry are further divided into a total of thirty-two symmetry classes, each with a different set of symmetry elements. All minerals fall into one of these thirty-two classes.

Calcite belongs to the trigonal subsystem of the hexagonal symmetry system, and to the trigonal-scalenohedral symmetry class. All minerals in this class have one 3-fold axis of rotation (the *c*-axis), three 2-fold axes of rotation (the *a*-axes), three mirror planes of symmetry (each perpendicular to one of the 2-fold axes) and an inversion center. The shapes and absolute sizes of the unit cells of minerals within the trigonal-scalenohedral class vary; however, the *a*-axes are always oriented at 120 degrees to one another and are perpendicular to the *c*-axis.

Aragonite belongs to the dipyramidal class of the orthorhombic symmetry system. All minerals in this class have three 2-fold axes of rotation (the *a*, *b* and *c*-axes) that are at right angles to one another. Each rotation axis has a mirror plane perpendicular to it. The unit cells of minerals within this class also vary in shape and size.

Fundamental Discoveries in Morphological Crystallography

The first major step in the development of crystallography was Nicolaus Steno's 1669 discovery, of the Law of Constancy of Interfacial Angles. Natural crystals are often distorted. Faces that are the symmetric equivalents of one another are often not the same size; thus, equivalent faces may have different numbers of corners and edges, and look quite unalike. The shapes of some crystals conform very closely to the geometric ideal, while others deviate strongly from it.

Steno demonstrated that while the faces may be different sizes and shapes, the angles between equivalent faces of a particular mineral are always the same.

Building on this and other observations, René Just Haüy and others gradually developed the above described concept of the unit cell.

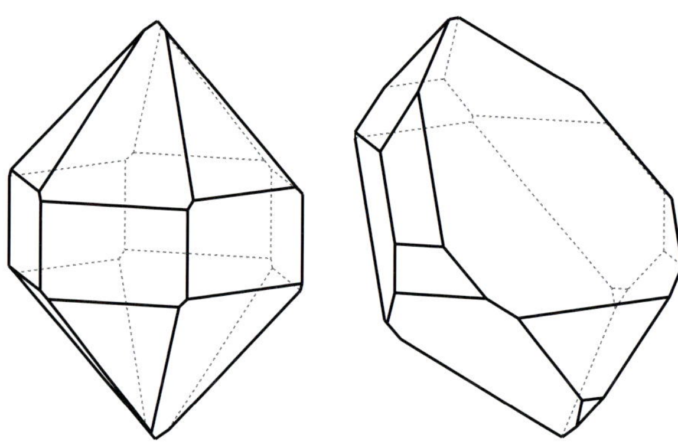

An ideal (left) and a distorted (right) quartz crystal are drawn in the same crystallographic orientation. Although the shapes of the crystals and many of their faces are quite different, the angles between equivalent faces on the two crystals are identical.

As techniques for measuring interfacial angles developed and more and more crystals were measured, crystallographers began to realize that the faces on crystals were related to the unit cell in a way formalized in the Law of Simple Rational Intercepts. This law states that crystal faces occur in orientations that are defined by whole (unit cell) steps along the symmetry axes, and that for a given face, these steps are equal multiples of whole numbers. Faces are not, therefore, randomly oriented to the crystal axes, but are confined to relatively few orientations.

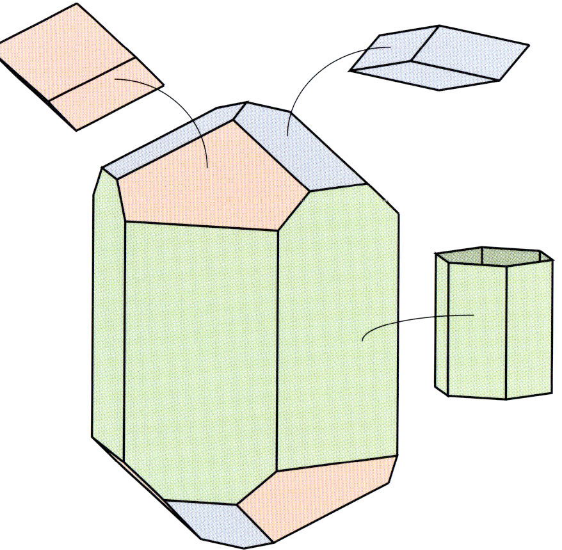

A calcite crystal that is composed of three forms: two rhombohedra and a prism. The smaller drawings show each form separately: the two rhombohedra $\{10\bar{1}1\}$ (left) and $\{01\bar{1}2\}$ (right) are shown above the central crystal; the prism $\{10\bar{1}0\}$ is on the middle right.

Miller Indices

Miller indices are the crystallographic standard used to indicate the points at which the planes of the unit cell intersect the crystal axes. Recall that these same axes are defined by the unit cell.

Suppose, for example, that a crystal face intersects the three axes at distances of 3, 2 and 4 units from the center of the crystal. To calculate the Miller indices, take the reciprocals of the intercepts: 1/3, 1/2 and 1/4. Clear the fractions by multiplying each by the least common multiple, in this case 12: 12/3, 12/2 and 12/4 which can be simplified to 4, 6 and 3. Finally, reduce the numbers by dividing each by any common factor; in this case the indices are already reduced. The Miller indices for the plane would be enclosed in parentheses and written without commas or spaces as (463). Faces that are parallel to an axis do not intersect it. The index for that axis is 0; thus, (100) is a form that intersects the first axis but is parallel to both the second and third axes.

Because calcite and other minerals in the hexagonal system have three *a*-axes, four axes in all, faces of calcite crystals are traditionally described using four indices. The third index is, however, geometrically redundant and always equal to the negative sum of the first two. For the above example, the Miller indices for a calcite crystal would be ($4\ 6\ \overline{10}\ 3$). Here we see two other conventions: when one or more of the indices has two digits, the indices are separated by spaces, and negative indices are written with a bar above the number, not a minus sign in front.

A direct consequence of the Law of Simple Rational Intercepts is that Miller indices are usually small integers, because the intercepts that define the Miller indices tend to be ratios of small integers. Miller indices such as (100), (110) and (111) are commonly seen in all minerals; indices such as (34 27 9) are most unusual and are considered suspect unless they are well documented.

Miller indices describe the orientation of a face relative to the crystallographic axes, not its absolute position; thus, a face that intersects the three axes at 312, 208 and 416 units has the same Miller indices as one that intersects the axes at 3, 2 and 4 (reduce 312, 208 and 416 by dividing each by the greatest common factor: 104). The shape of the crystal and its faces is what is important, not their sizes.

There are subtle differences in the several ways that Miller indices are used to describe crystals. For example, indices that are enclosed in parentheses, such as ($10\overline{1}0$), refer to the orientation of a given face; indices enclosed in brackets, as {$10\overline{1}0$}, refer to a form and express all of the faces on a crystal that are symmetrically equivalent to one another.

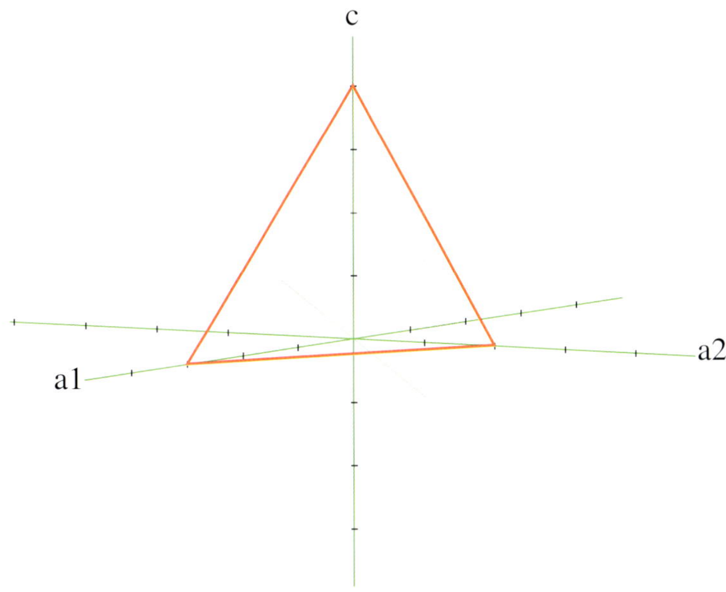

The natural coordinate system for calcite showing, in red, a crystal face that intersects the a1, a2 and c-axes at distances of 3, 2 and 4 units respectively. The Miller indices for this face are (463).

Predictability and Variety

All calcite crystals have the same unit cell, the same symmetry, the same atoms and the same bonds, so why don't all calcite crystals look alike? The reason, in a word, is *environment*.

In a single vug, or even an entire locality, calcite crystals are usually composed of the same forms and have about the same shapes. But from place to place and sometimes even from vug to vug, there is great variety.

There is still much to learn about the factors that determine the form of a crystal in a specific location; however, observations and experiments point to many factors in the genetic environment that influence the form of the resulting crystals, including rate of growth, temperature and impurities in the growth environment. These factors interact to determine the forms that make up a crystal and the prominence of one form relative to another.

Within a basic plan controlled by crystal structure and resulting symmetry, there is ample room for variation to reflect the individual character of different genetic environments. Of all minerals, calcite is the best example of this variation within a basic plan.

Calcite Pseudomorphs

***Rupert Hochleitner** on calcite that has disappeared but not without a trace*

Have you ever found a specimen, a scalenohedron or rhombohedron completely typical of calcite, but realized on closer examination that the crystal was hollow, was quite hard or did not effervesce in acid? What you have likely found is not calcite at all, but a pseudomorph after calcite.

Pseudomorphs are particularly interesting crystal forms with the potential to provoke thought and debate, as the circumstances surrounding their formation often prove mystifying.

On a basic level the concept of a pseudomorph is simple. A pseudomorph is a nicely formed crystal whose molecules have been replaced by those of another mineral. This replacement has occurred so precisely that the original crystal's shape has been retained; thus, the newly crystallized mineral has a form that is not its own.

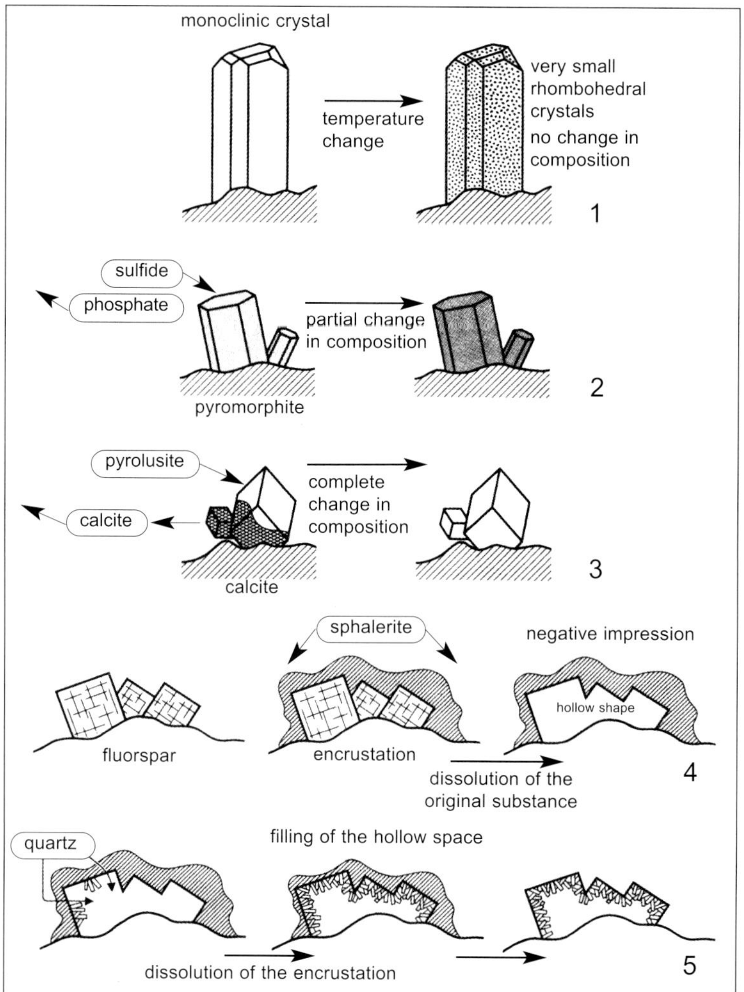

Formation of pseudomorphs (schematic)

1) Paramorph of a-sulfur after b-sulfur
2) Alteration pseudomorph of galena after pyromorphite
3) Replacement pseudomorph of pyrolusite after calcite
4) Encrustation pseudomorph of sphalerite after fluorite
5) Cast pseudomorph of quartz after fluorite

after Lieber, 1969

False Form

Pseudo is a Greek term meaning *false*, *morphe* means *form*; a pseudomorph is a false form. Having their origins in replacement, pseudomorphs usually lack the direct connection between internal structure and external form that typifies *normal* mineral crystals.

In most cases, pseudomorphs are composed of many small crystals, although their form is that of a single larger crystal. In rare cases, the replacement may also be a single crystal, but there is generally no relationship between the replacement mineral (identified for example by cleavage) and the shape that it inherited from the original mineral. The disconnect between external form and internal structure is one way of recognizing a pseudomorph.

While a *normal* mineral specimen exhibits its true form, a pseudomorph also reveals something of its past. From it, one can learn something about the geologic conditions surrounding a specimen's formation through at least three distinct stages: the period in which the original crystal formed, that in which the original crystal was replaced and sometimes a third period in which the final form took shape. Pseudomorphs are thus scientifically important, as they permit a view of the local conditions over a protracted period.

Pseudomorph Classification

In his 1969 book *Mineralogie in Stichworten* (Abbreviated Mineralogy), Werner Lieber classified pseudomorphs into five types.

1. A **paramorph**, or rearrangement pseudomorph, is a solid form in which a mineral exhibits the shape of one of its polymorphs. In the case of calcium carbonate, three low-temperature polymorphs exist: trigonal calcite, orthorhombic aragonite and hexagonal vaterite. Among these three, the trigonal modification

Pseudomorphs After Calcite

(calcite) is the most stable; thus, paramorphs of calcite after aragonite are more frequently found than those of aragonite after calcite, which are quite rare. Vaterite is uncommonly observed, as it is the least stable of the polymorphs and typically reverts to aragonite or calcite.

2. An **alteration pseudomorph** is a solid crystal in which the chemical composition of the original mineral was partially altered, rather than entirely dissolved, to form the new mineral, and some element of the original mineral is retained. An example (lower right) is rhodochrosite ($MnCO_3$) after calcite ($CaCO_3$), in which the carbonate molecules from the calcite are incorporated into the new rhodochrosite crystals.

3. A **replacement pseudomorph** is a solid form in which the original mineral was completely replaced by a mineral with which it has no elements in common. One example (upper right) is pyrolusite ($Mn^{4+}O_2$) after calcite ($CaCO_3$).

4. An **encrustation pseudomorph** is a hollow form that occurs when a crystal becomes encrusted with another mineral, after which the original crystal dissolved. What remains is a shell in the shape of the original crystal. This negative crystal is referred to as a *mold.*

Because it is soluble in every acid, calcite is often the original crystal in an encrustation pseudomorph. Certainly, the most common encrustation mineral is quartz, but others such as dolomite, fluorite and hematite also form encrustation pseudomorphs. The specimen of pyrite after calcite pictured on the right is an encrustation pseudomorph.

5. A **cast** is a two-step pseudomorph in which the negative void space of an encrustation pseudomorph is *partially or entirely* filled by another (or the same) mineral. The encrustation material may remain or may also dissolve, leaving only the cast. If the encrustation material has dissolved, it can be difficult or impossible to distinguish a cast from a replacement pseudomorph.

Calcite Pseudomorphs

Calcite occurs as the original material of a pseudomorph with relative frequency. While it is more or less common for other minerals to

Beauty and the Beast

*Sure to excite the "black-ugly" collector in everyone, this 8 cm wide **replacement pseudomorph** of pyrolusite after calcite from Ilfed in the Harz Mountains of Germany is part of the Bavarian state collection.*

Photo Rupert Hochleitner

*Easier on the eye, this 2 cm wide **encrustation pseudomorph** of pyrite after calcite is from Rensselaer Stone quarry in Pleasant Ridge, Indiana.*
Harris Precht collection; Terry Huizing photo

A Mnemonic

The description of a pseudomorph follows a specific pattern: it is always a pseudomorph of mineral A after mineral B, in which mineral A is the one presently observed, while mineral B is the original material, generally only recognized by form. A pseudomorph is simply described as, "What is after what was."

Special thanks to Dave Ellis for his critical review of and helpful suggestions toward the accuracy of this article

Transformation: Rearranging, Altering,

An encrustation pseudomorph of dolomite after calcite from Grevenbroich in Sauerland, Germany

Above: *The original calcite shape is evident; field of view 12 cm*

Below: *Inside the mold, the calcite that dictated the specimen's shape is missing, having been replaced by dolomite*

Bavarian state collection; Rupert Hochleitner photos

replace calcite, it is uncommon for calcite to replace another mineral.

As calcite is highly soluble, replacement pseudomorphs, encrustation pseudomorphs and casts are much more common than paramorphs or alteration pseudomorphs involving calcite.

Pseudomorphs are not confined to particular types of deposits, though they are more common in some types of deposits than others. Calcite pseudomorphs are rarely found in limestone fissures, which often have a simple history that favors the growth of calcite but not its dissolution. Calcite pseudomorphs more often occur in hydrothermal deposits and oxidation zones, which often have complex histories that may favor the growth of calcite at one stage, and its dissolution at a later stage. For example, some of the fluorite veins found at Wölsendorf in Bavaria, Germany are extremely rich in calcite pseudomorphs.

Some authors refer to so-called specimens of *sand calcite* (lower left photo page 25) as a pseudomorph; however, these are normal calcite crystals with sand grains as inclusions in the crystal structure. As with many aspects of field identification recognizing and classifying pseudomorphs requires practice.

Taking a Closer Look

The encrustation pseudomorph of dolomite after calcite, pictured on the left, began as a typical calcite scalenohedron, built of numerous tiny rhombohedra. A hydrochloric acid test immediately reveals the fact that the present mineral is dolomite, and observing the surface of the scalenohedron with a loupe, individual dolomite crystals are easily recognizable. From the underside (lower left), the scalenohedron is hollow; thus, the specimen is clearly an encrustation pseudomorph.

The inside of the hollow scalenohedron is a mold of the surface of the original calcite crystal, but the inner surface of the mold is not smooth, as might be expected. Instead, webs of dolomite crystals, all with precisely the same angle between them, have grown toward the interior of the mold.

Replacing, Encrusting and Filling

Interestingly, the angles between the dolomite crystals are the same as the angle between two cleavage planes of calcite. This effect results from the fact that calcite, when etched, is not etched uniformly. In spots where cleavage cracks were present on the scalenohedron, the etching solution penetrated the crystal more quickly, encrusting those areas more deeply than the clean surfaces. Additional, tiny dolomite crystals formed along the cracks creating webs.

The foundation had been laid for the encrusted calcite to be slowly altered to dolomite; however, the process of dolomite deposition stopped before the scalenohedron had been replaced (altered). The original calcite entirely dissolved, leaving a hollow space in which the dolomite deposited by the encrustation solution is still visible.

Pseudomorphs From the Sea Floor

California pseudomorph collector Keith Harshbarger relays the history of an intriguing example of alteration pseudomorphism: calcite ($CaCO_3$) after ikaite ($CaCO_3$ 6 H_2O), pictured on the lower right.

Ikaite is a monoclinic, calcium carbonate hexahydrate that is stable up to about 3° C; it seems to form at the sea floor near bicarbonate springs. Granular ikaite was first discovered by a Danish expedition working the Ika Fjord, on the southwestern coast of Greenland. Pauly named the new mineral in 1963 after the discovery site. Above temperatures of 3° C, ikaite rapidly loses water disintegrating into porous, granular calcite with a monoclinic shape.

In 1982, the crew of a German ship collected clear, glassy ikaite crystals to 6.5 cm long from sedimentary cores of the sea floor off the coast of Antarctica. These crystals soon turned opaque and disintegrated into granular calcite and water. Luckily, a photo was taken of the original crystals, and the connection was later made between these specimens and the *pseudo gaylussites*, *thinolites* and other previously unknown pseudomorphs found around the world, typically in glacial lake sediments or in sedimentary sea floor deposits originating near the Arctic Circle.

Above: *The original calcite crystal provided both its shape and carbonate molecules to this 7.8 cm wide* ***alteration pseudomorph*** *of rhodochrosite after calcite (yes, the crystal is solid) from the Idarado mine near Telluride in San Miguel County, Colorado. Collection Dave Bunk; photo Jeff Scovil*

Below: *This 10.2 cm wide example of calcite after ikaite was found at Bielo More on the Kola Peninsula in Russia. Martin Zinn collection; Jeff Scovil photo*

"What Is After What Was"

The various colors of these pseudomorphs are likely a function of the composition of the muds in which the crystals grow. These muds almost certainly fill the voids that opened as the original ikaite crystal dehydrated.

An encrustation pseudomorph of hematite after calcite from Sandwig bei Iserlohn in Westphalia, Germany; the largest crystal is 4 cm long; Terry Huizing collection and photo

Pseudomorph After Pseudomorph

Rearrangement, alteration and replacement pseudomorphs as well as casts are all pseudomorphs that result in solid crystals: they are never hollow. It is thus fairly easy to identify an encrustation pseudomorph as it is the only type of pseudomorph that results in a mold.

Over the course of their geneses, some crystals begin as one type of pseudomorph and end as another. The pseudomorph of hematite after calcite depicted above was formerly in the collection of Scotsman Robert Ferguson (1777-1846).

Geochemist and pseudomorph collector Dave Ellis postulates that it enjoyed a rather complex history. He suspects that the calcite crystals were first encrusted by another mineral such as gypsum. The calcite dissolved leaving a mold of the original calcite crystals. Hematite filled the calcite-shaped molds and encrusted the mineral that had originally encrusted the calcite. The hollow *bubbles* about the surface of the specimen were likely left when the first encrusted mineral dissolved from beneath the encrusting hematite. Terry Huizing currently owns this specimen, and he indicates that the calcite crystals are hollow. In spite of its history, this specimen, is thus simply an encrustation pseudomorph of hematite after calcite.

Polyhedral Agate

Brazil has produced specimens that appear, at first glance, to be the agate variety of quartz filling the negative void of an encrustation pseudomorph, a classic cast pseudomorph. Sliced open, these samples have the appearance of agate geodes with amazingly straight edges (facing page, lower right) and flat surfaces. Collectors theorized that the original molds were made after calcite crystals. Uncut samples, however, evidenced shapes inconsistent with the crystallography of calcite, or any other mineral for that matter. The faces of these casts had a random configuration that varied widely from specimen to specimen, consistent only in having flat surfaces. The surfaces had to have been shaped by a crystal, but how?

In 1982, Andreas Günther published the accepted solution to the puzzle. He suggested that the quartz had formed as an encrustation of groups of calcite crystals. Quartz lined the voids between calcite crystals and different adjacent crystals and shaped the *faces* of the forming quartz crystal.

Interestingly, some agate-filled geodes have a hexagonal void at their center. These voids were shaped as the geode filled in around a calcite crystal, lending its shape to the geode's center.

While they are certainly false forms, polyhedral agate specimens are not widely classified as pseudomorphs.

False Forms

Shaping the Fossil Record

Objects other than crystals also lend their shapes to subsequent mineralizations. Organic matter and fossils are subject to encrustation, alteration, replacement or casting in a manner similar to the pseudomorphing of crystals.

Silica, the oxides, phosphates and carbonates are among a number of mineral groups that are known to fossilize the shells, bones and teeth of many of the species included in the fossil record.

This 15 cm wide cross section of a fossilized ammonite is from Belorechensk in North Kaukaz, Adigeja, Russia. The original structures were replaced by calcite and the voids filled with calcite and aragonite. Photo Jeff Scovil

Pseudomorphs?

Left: *In spite of reports to the contrary, sand-included calcite, though lovely, is not a pseudomorph. This 5 cm wide specimen is from Cabreret in Lot, France. Collection and photo Terry Huizing*

Right: *Taking their shapes from the voids between crystals, polyhedral agates are generally not considered pseudomorphs. This unusually equant 9.4 cm wide cross section is from Sitio Garuelo, Municipe de Cachoeira dos Indios in Paraíba, Brazil. Louis-Dominique Bayle collection; photo Jeff Scovil*

René Just Haüy

Cleavage and

Lydie Touret investigates a famous old story: the broken calcite crystal of René Just Haüy (1743-1828).

Lydie Touret is the curator of the Musée de Minéralogie at the École des Mines in Paris, the same institute at which Haüy worked about 200 years ago.

The popular consciousness tends to latch onto moments of genius to explain great discoveries: Archimedes shouting, "Eureka!" in his bathroom or Newton sitting under an apple tree in Cambridge when an apple falls on his head.

Similarly for the last 200 years, mineralogy textbooks have recounted the anecdote of the calcite specimen that slipped out of the clumsy hands of the famous Abbé Haüy. The calcite smashed into rhombohedron-shaped pieces, and the meaning and causes of cleavage shot like lightning through the head of the genius. The tale next has Haüy picking up the broken pieces and trying to cleave them again. Noticing that they all had the same angle and parallel surfaces no matter how small they got, Haüy imagined breaking them into the tiniest unit, the *molécule intégrante*, a building block that permitted the most diverse crystal shapes to be built when stacked in all three dimensions. Modern crystallography, the science of the structure and symmetry of crystals, was thus born!

Haüy Defines *Mineral*

As is so often the case, the truth is not nearly as simple as the story implies. A lengthy maturation process generally precedes most discoveries, though sometimes the process is a subconscious one. While actual events may not be as romantic as the story has become, Haüy undoubtedly did a great service. We are indebted to him for the first good definition of what constitutes a mineral species and for the discovery and description of many minerals and rocks including pyroxene, dioptase, kyanite, orthoclase, eclogite and pegmatite.

According to Haüy, a mineral is a natural solid of homogenous chemical composition with a regular structure that, by geometric transformation, forms crystals bounded by flat faces. The definition has been refined over time, but Haüy's concept has led most mineralogists, even the Russians (who customarily have their own discoverers for everything), to consider Haüy the father of modern mineralogy. Haüy's 1801 *Traité de Minéralogie* became the standard work of its time. First translated into German by D.L.G. Karsten and C.S. Weiss in 1804, the work has never been translated into English.

Indeed, Haüy's concept of the *molécule intégrante* was an important step in solving the question of the structure of matter; nevertheless, the picture of the modest, selfless man of learning that was presented by the biographers of the time, and by Haüy in his own works, does not express the whole truth. The present text is probably not the forum for rehashing the bitter controversies that he had with the advocates of isomorphism (variable chemical composition in the same crystal habit), prominently with the German chemist and crystallographer Eilhard Mitscherlich (1794-1863). When all is said and done, very few people really doubted Haüy's *molécule intégrante*.

The Story of the Broken Calcite

One of the many versions of the anecdote simply tells of a dropped crystal. In others, the story is not one of clumsiness, but of an intentional scientific experiment focused on further developing the concept of crystal structure. Haüy said that his breaking the crystal was a "key to the theory." In his *Traité de Minéralogie*, Haüy wrote, "It (the key to the theory) impressed itself on me when 'Citoyen Defrance' kindly gave me a crystal from his mineral cabinet that he had just broken out of a vug. The prism had a broken surface at the terminal edge of the basal pinacoid at which the crystal terminated. Instead of putting the crystal into my collection, which was then being formed, I tried to split it in another direction; and after several attempts, undertaken in uncertainty, I succeeded in pulling out its rhombohedral nucleus. The surprise I felt at this was equally tied to hope that this first step would not be the end of the subject."

Was this discovery really as fortuitous as the shattered calcite? Haüy himself always claimed so, but there are many indications of the influence, perhaps subconscious, of his predecessors. The idea seemed to be hanging in the air. Worse yet, others had previously published the idea, a fact that casts doubts on Abbé Haüy's integrity. A digression into the history that preceded Haüy's discovery may shed light on the matter.

the Birth of Modern Mineralogy

From Building Stones to the Microscope

From ancient times until the 17th century, knowledge about minerals and crystals had developed slowly, but beginning in the Middle Ages, greater strides in mineralogy were stimulated by the extraction of ore deposits as well as by the preparation of building stones for forts and cathedrals. The science's rapid development led to important observations. There was a good deal of interest in splitting stones in favored directions. In order to facilitate some of the more strenuous tasks in mines, quarries and all sorts of workshops, it was a great advantage to know in detail how rocks would cleave.

The various schists and slates are easily split; thus, they were the ideal material for the roofs of buildings. The picture plates in **Diderot and D'Alembert**'s 1765 *Encyclopedia* provide an image of the interest, at the time, in the various ways to split stone. No less than twelve different directions were observed in some stones, and each direction was referred to by a unique term.

The Moment of Genius

One day in the year 1780 – the exact date is unknown – it happened: Haüy dropped a calcite on the floor and it broke into pieces that were all the same! In them he saw the solution to the puzzle…

Scenes from the 1993 documentary film about Haüy by Lydie Touret

René Just Haüy

Some of the terms were still being used in the beginning of the 20th century but have since been lost. Interestingly, the descriptions in the *Encyclopedia* are included under the title *Mineralogy*.

Over time, these observations became more refined, and attention was paid to smaller and smaller units until individual crystals finally began to be observed. Of course, there was no concept, at this point, of homogeneous structure or of minerals in the modern sense.

In the early 16th century, knowledge of crystals slowly began to flower. Observations became ever more precise, and observational tools, most importantly the microscope, developed steadily. While looking through one of the first microscopes, **Anton van Leeuwenhoek** of Delft (1632-1723), discovered red blood cells. He was also interested in crystals, and it was during the same time that the first morphological observations of crystals were made; the cubes of salt were differentiated from the six-sided prisms of saltpeter and the octahedrons of alum on the basis of their crystal habit.

Calcite's cleavage was soon observed, in large part because the property is so easily seen. Calcite cleavage also lent itself to simple experiments: calcite, unlike quartz, can easily be cut with a knife into perfect sections.

The 17th century saw the development of more than the microscope. Dutch academic **Christian Huygens** (1629-1695), a friend of Leeuwenhock, formulated the basic laws of optics. After discovering the laws of the refraction of light, he occupied himself with the birefringence (double refraction) of calcite. His material of choice was, of course, *Iceland spar*, marvelous crystals that are completely clear. Danish scientist **Erasmus Bartholinus** (1625-1698) was the first to observe double refraction in calcite (see page 40). The interpretation of this phenomenon is a significant element in Huygens' 1660 *Traité de la Lumière*.

We extend our thanks to Yves Charlesbois for lending his knowledge of the English and French languages to the accurate translation of Haüy's original writings.

The First Angle

At the same time, crystal forms and their relationship to cleavage were also attracting the attention of other learned people. Bartholinus reported that fragments obtained by cleaving Iceland spar, which in those days was still available in quantity, were always identical in shape to the original crystal!

Nicolaus Steno (1638-1686), also a Dane, recognized the significance of the angle between the crystal faces. He found the angles between adjacent edges of the rhombohedron to be 101° and 79° and the angle between adjacent faces to be 103° 40'. These measurements were refined by Huygens to 101° 52', 78° 8' and 105°, respectively. Huygens also made the fundamental observation that the cleavage always runs parallel to the rhombohedron faces!

The geometric analysis of crystal and cleavage forms foreshadowed the modern approach to the topic, but the historical interpretation of the data was strange indeed. Without the slightest idea about the chemical composition of minerals, seventeenth century scientists classified mineral species according to their physical properties (density, hardness, etc.) as well as by the ease and perfection of their cleavability.

Calcite: the *Iceland Talc*

Gypsum, the most easily cleavable mineral, was in those times referred to as *talcum*; thus, a controversy arose as to whether calcite should be considered a relative of gypsum or of other minerals like quartz. Bartholinus chose the latter, but Huygens contradicted him, "One must suppose that it is a kind of talc and not a kind of crystal (*crystal* is a term that was formerly applied to all hard, glassy minerals, especially quartz)."

Bartholinus' main opponent was the Frenchman **Philippe de la Hire** (1640-1718), who had just developed a new mineralogical classification based on cleavage. Considering calcite's double refraction, which he could not find in quartz, and considering the perfect cleavage of calcite, de la Hire proposed that all easily cleavable species be grouped together as *talc*. He said "not without reason can it (calcite) be better designated as talc than as quartz, since its most prominent property is easy cleavage." He, therefore, referred to calcite as *Iceland talc* and expanded its definition to cover gypsum, his preferred experimental material. Gypsum was common in the building stone quarries around Paris where de la Hire lived.

The observations on gypsum, which has a perfect and easy cleavage, were generalized and applied to calcite. These generalizations represented a

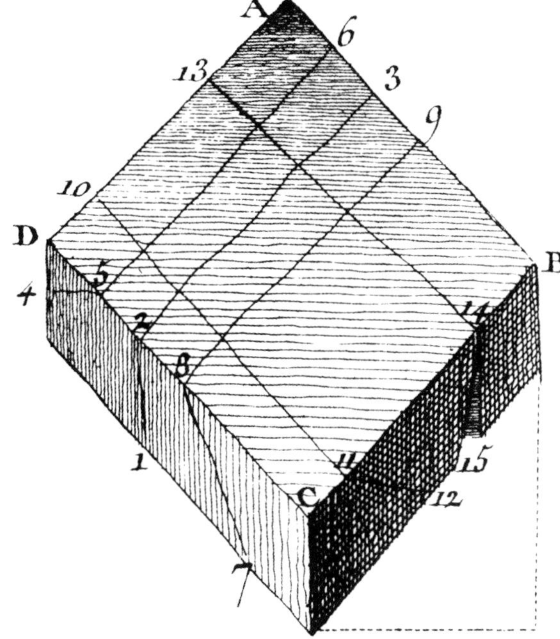

First Observations of Cleavability

These drawings were taken from the Encyclopedia *of d'Alembert and Diderot, 1765*

***Above:** A slate layer in a quarry near Rimogne in the Ardennes*

***Below:** The individual directions in a block of slate were described as East, West, South and North, and by different qualities.*

step backwards from the progress made by Bartholinus and Huygens. It was not until the angles were more precisely measured that it became possible to explain the elements of symmetry and the fundamental principles of crystal structure. Although discoveries about the nature of cleavage were significant in building the foundation of our understanding of crystallography, cleavage by itself was not enough to lead to an understanding of the structure of rocks or of crystals.

Preparing the Way for Haüy

It was thus established before Haüy that cleavage was distinct in some minerals and less distinct in others. Haüy's fundamental discovery was that there could be more than one direction of cleavage and that each was parallel to the *molécule intégrante*.

Haüy's work hardly would have been possible without the angle measurements of Steno, not to mention those of French crystallographer **Jean-Baptiste Louis Romé de l'Isle** (1736-1790) and his coworkers. Romé used a contact goniometer known as a *mesure-angle* that was developed by his assistant **Arnould Carangeot** (1742-1806).

On almost all of his portraits, Haüy was depicted with such a goniometer placed on a cleavage rhombohedron. Remaining loyal to this instrument for his entire life, Haüy refused to accept the significantly more accurate measurements of **William Hyde Wollaston**'s (1766-1828) new reflection goniometer.

Making countless measurements on clay crystal models, Carangeot empirically discovered the *law of constancy of angles*, which states that the faces of a natural crystal can be of different sizes, but that the interfacial angle remains exactly the same. This law, known in older French literature as the *Law of Romé de l'Isle*, had been formulated by Steno for quartz more than one hundred years earlier, but Romé and Carangeot extended the observation to all minerals. Carangeot is known to have been unaware of Steno's work during the time at which he made his observations.

This law is a significant building block in the edifice of Haüy's theory, but in his papers he refrained from admitting his debt to Romé and Carangeot. In fact, the relationship between Haüy and Romé was turbulent from the start, though Romé seems to have been the more aggressive of the two.

René Just Haüy

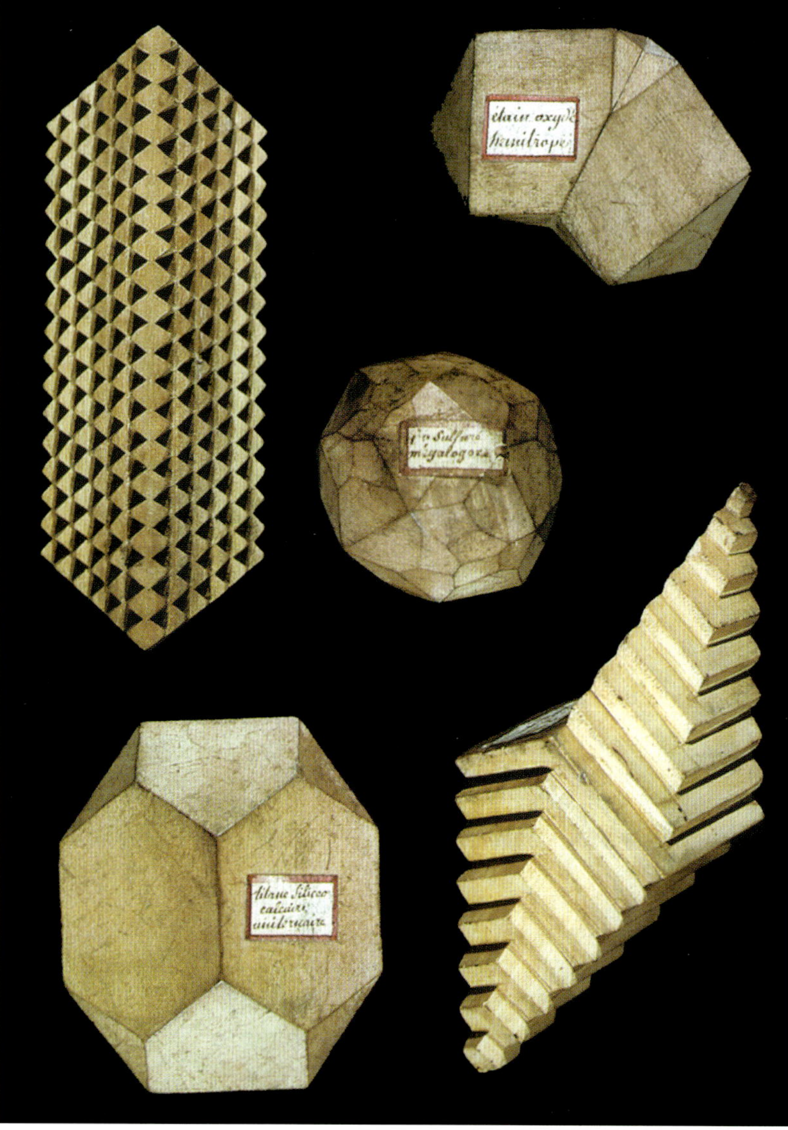

Haüy's Original Models

These served as his study models. The model on the lower right shows how a calcite scalenohedron is built of many tiny rhombohedra (according to Haüy's "decrescence theory"). Photo by Peter Stemvers

Enter Haüy

In 1781 while at the Collège Cardinal le Moine, Haüy presented his first two papers at the *Académie des Sciences* in Paris: one on February 21st on garnet, the other on December 22nd on calcite. In both papers, Haüy reported on the relationship between cleavage and crystal structure. His presentation was met with excited acceptance at the Academy and drew attention to this young scientist.

Both papers dealt with the basics of cleavage and were published 1782 under "Observation sur la Physique, sur l'Histoire Naturelle et les Arts " in the May/June issue of *Journal de Physique* (Haüy, 1782). One important excerpt: "I (Haüy) tried to investigate the (structure of crystals) by creating sections (cleavages) with a cutting tool. They are a sure indication of the structure, since they can only be carried out in certain directions. The cleavages look like they have been cut. All breaks in other directions produced irregularly shaped fracture surfaces."

Calcite is an obvious choice for these experiments, but the subject of the first paper is quite shocking. Garnet has no cleavage directions, and it is quite impossible to cleave pieces off "with a cutting tool." Haüy unconvincingly claimed that cleavable minerals often show fractures and parallel striations. He asserted that some minerals are too difficult to cleave "in smooth slices," but still have external striations, as is the case with garnet. Haüy concluded that these striations, fractures and other superficial indicators evidenced cleavage.

"Crystal Smasher"

Haüy and his theories were attacked by Romé de l'Isle, whose position was backed by mathematicians de la Place and Bezout. Romé dismissively referred to Haüy as the "Cristalloclaste", the *Crystal Smasher*!

After his papers were presented, Haüy countered the attacks by outlining his technique. Haüy reported that he broke the crystals by throwing them into boiling water and sharpened his powers of observation by looking at the broken sections through a hole in a card.

Dropping a stone was more than just clumsiness in those days. That point is illustrated by an anecdote about mineralogist **Abraham Gottlob Werner** (1749-1817) of Freiberg. In 1897, science historian Abbé J. Loridan wrote, "People tell of another famous mineralogist, Werner, who could not tolerate anyone's handling minerals clumsily. Even the slightest attack on their sharpness and luster wounded him deeply. He once said about a colleague, 'He may well be a great minister and a talented general but,' he sighed, 'He doesn't even know how to hold a mineral.'"

An even more serious accusation, however, was brought by Romé in his 1783 book *Cristallographie*. Romé wrote that Haüy's calcite thesis was copied from the writings of Swedish mineralogist **Torbern Olaf Bergman** (1735-1784), who had already said that "all calcite crystals have a rhombohedral core, just like *Iceland spar*." Romé wrote, "I know very well that, imitating the famous Torbern Bergman, some scientists among us today occupy themselves with figures and geometrical calculations to explain the special structural mechanism of some crystals which are easily split with the help of a cutting instrument... now Monsieur l'Abbé believes he can assert that calcite crystals have a rhombohedral core, the same as Iceland crystal, and that it is possible to expose it by taking away the excess material layer by layer." This was a serious accusation of plagiarism!

Haüy must have been deeply affected by this charge, as he had based his entire theory on this observation. Be that as it may, Bergman had previously described two essential elements of cleavage and calcite. The accusations instigated some rather confused responses on Haüy's part, and looking more carefully at Bergman's statements, there do indeed appear to be a series of perplexing similarities. Without using the term *molécule intégrante*, Bergman explains the diverse habits of calcite as deriving from a *rhombohedral core*, which he regarded as the archetype. Bergman's published drawings are unassailable. He even extrapolated his theory, still with the same rhombohedral core, to garnet (!), then corundum, pyrite, tourmaline and staurolite. In his theory, Bergman made mistakes, such as confusing single crystals with twins. Haüy later corrected these errors.

In a striking coincidence, Bergman explained that he owed his ideas to an accidental discovery by his student Johann Gottlieb Gahn, who dropped a *pig's tooth* (scalenohedral) calcite crystal that broke into small rhombohedral fragments!

The rhombohedral core, the broken crystal, calcite and garnet as the first investigated minerals, Romé seems to have had serious grounds for his accusation.

Haüy later addressed the issue himself, but in a manner that did nothing to allay any doubts. In his 1782 report on calcite to the Academy he does mention Bergman, but only to distance himself from him, "The structure of a crystal made up of twelve five-cornered faces, as developed by me, is very different from that supposed by Bergman in his 1779 treatise." Bergman's treatise is not dated 1779, which would have made it almost contemporaneous with Haüy's 1782 paper, but rather 1773. It is thus nine years older than Haüy's: old enough to have become known among the small circle of mineralogists of those days who were already attentive to whatever came from the Swedish school.

To Romé's accusation of plagiarism, Haüy answered very briefly in a 1784 essay, "At the time when I was beginning the study of crystals, I had the opportunity to read a paper on crystallization by Bergman, who had been in

René Just Haüy 1743 - 1822

The important Abbé Haüy always had his portrait painted with a calcite cleavage rhombohedron and a goniometer.

Oil painting by André Nicolas Van Gorp

Uppsala since 1779." Again the same mistaken date, but more importantly no analysis of Bergman's ideas, as if it were distressing for Haüy to even broach the subject.

Even more perplexing is that in 1801 Haüy, probably assuaged by the absence of Romé who died in 1790, did not hesitate to deny any possible influence on the creation of his work. "The Academy already had knowledge of my first papers when they received Bergman's treatise; I was told of it because it could have been of interest to me, since it was related to my work."

It is not very likely that a paper that had been prepared in 1773 was not presented to the

René Just Haüy

Contact Goniometer Designed by Carangeot

Above: *The hinge at the 90 degree mark, allowed the left half to be folded away; 15 cm diameter, signed W. Eckling, Vienna, about 1830; Ulrich Burchard photo*

Right: *René Just Haüy*

Academy until 1781; thus begging the question as to whether this repeated mistaken date was merely coincidental. Haüy later made an effort to play down Bergman's role and stated in his 1801 textbook, "He (Bergman) was satisfied with his first impressions and concerned himself neither with developing structural laws nor with quantifying the relationship involved; he merely provided a sketch, outside the perspective of mineralogy, while revealing the skill of his hand that contributed with much success to the science of chemistry."

Birth of the New Mineralogy

Almost all of the authors occupied with the history of mineralogy – Marx 1825, Groth 1826, Mauguin and Orcel 1944 and many others – shared the opinion that the discovery of the *rhombohedral core* in calcite ushered in the birth of a new science. Haüy is unanimously celebrated as the great man of this period; his fame is unlikely to fade.

Although the real Haüy may not have fit the ideal persona that history has afforded him, he still contributed a great deal. He wrote numerous publications, maintained contacts with many learned individuals and societies, built collections of crystal models and much more.

While Haüy never gave credit to his predecessors for their obvious influence, no great spirit is without weakness. The science historian R. Hooykaas, who undertook a thorough study of the beginnings of Haüy's crystallography as revealed by the original documents, strikes a good balance. "It is perhaps a greater contribution to build an exact theory from beginnings, mistakes and half-truths, whether those of others or one's own, than to create it from nothing." (Hooykaas, 1955)

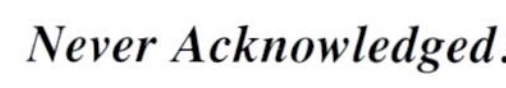

Never Acknowledged!

Left: *Bergman's concept of the "rhombohedral core" of calcite, published 1773*

Right: *Haüy's analogous drawings of 1781; the inserted rhombohedra depict the position of the main rhombohedron for the whole crystal*

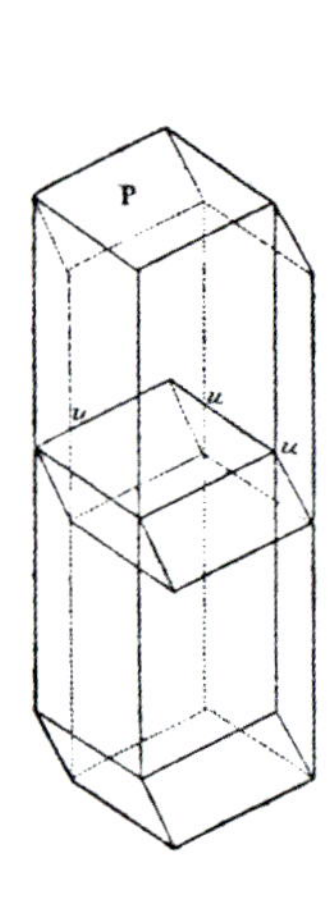

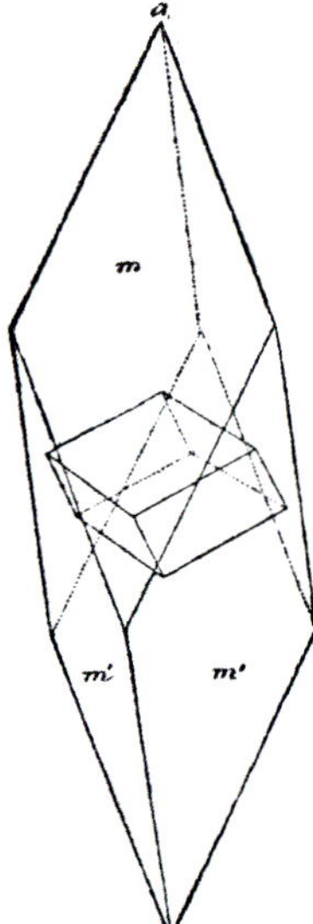

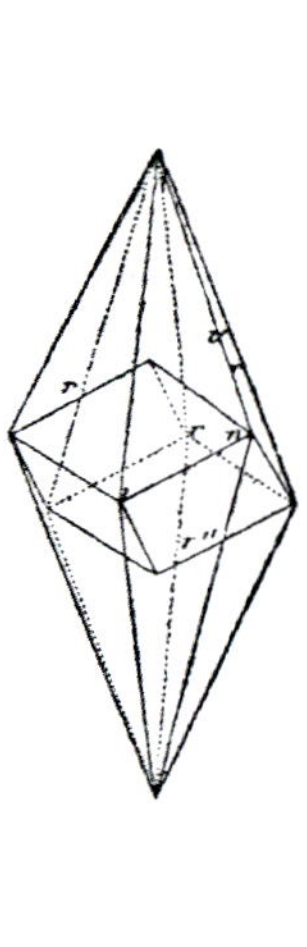

Cleavage

by Pete Richards

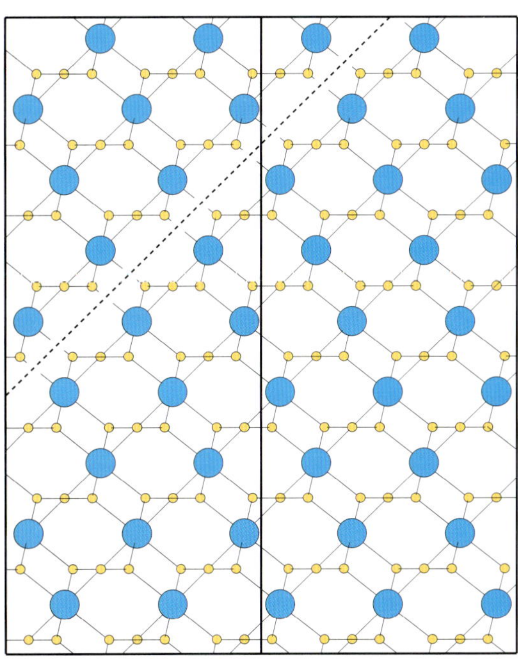

Diagram of a prismatic calcite crystal that shows the crystal structure; calcium atoms are represented by blue circles, oxygen atoms by yellow circles. In this diagram, the c-axis is vertical, and the view is down an a-axis that is perpendicular to the page. For any crystal that the human eye can see, the pattern of atoms indicated by this diagram extends at least a million times in all directions. Drawing by Pete Richards

Crystals are aggregates of atoms that are arranged in a regular pattern and repeated in three dimensions. In a crystal, atoms are connected to one another by chemical bonds. The strength of these bonds, and thus the strength of the overall crystal, depends primarily on the kinds of atoms that the bonds connect.

Because crystals are made up of repeated patterns of atoms (and bonds), those minerals that have weak bonds in their structures also have planes of weakness extending through them. These weak planes correspond to the direction in which the weak bonds are repeated in the structure. When a crystal breaks, it tends to break first along any weak planes creating a flat surface or *cleavage*, which for a given mineral always occurs in the same orientation relative to the crystal's axes of symmetry.

Cleavage is described by terms such as *highly perfect*, *perfect*, *good* or *difficult* to reflect how consistently a given mineral tends to break along these systematically oriented planar surfaces. Examples of minerals that exhibit each of the above four classes of cleavage are mica, calcite, barite and quartz respectively. Breaks that do not occur systematically along cleavage planes are referred to as *fractures*. There is a terminology for fractures that is analogous in its complexity to that for cleavages, but it is adequate to mention that the most typical fracture is *conchoidal* (i.e., fractured along a curved surface).

Minerals that do not have cleavage show some form of fracture when they are broken. Minerals that have only one or two directions of prominent cleavage show some form of fracture when broken in non-cleavage directions. Quartz is generally regarded as having very poor or no cleavage. It typically breaks to show a conchoidal fracture.

The quintessential example of a mineral with cleavage is mica. Micas are made up of planar, aluminosilicate sheets that are composed of strongly-bonded silicon, aluminum, oxygen and hydrogen atoms. These sheets are separated by layers containing large positively charged atoms, such as potassium or sodium, that are weakly bonded to the aluminosilicate sheets. Micas show a single direction of highly perfect cleavage. This cleavage is the result of the contrast between the weak bonds that connect the positively charged atoms to the aluminosilicate sheets and the much stronger bonds within the aluminosilicate sheets. A stack of paper, with its sheets weakly held together by static electricity, is a reasonable analogy. The sheets of paper correspond to the aluminosilicate layers, and the static electric charge corresponds to the weak charges that hold these layers together.

Calcite has three equivalent directions of perfect cleavage, reflecting its threefold symmetry. Each cleavage direction is inclined at the same angle to the *c*-axis.

Calcite crystals consist of layers of carbonate groups, alternating along the *c*-axis with layers of calcium atoms. Each carbonate group consists of a carbon atom that is surrounded by a triangle of oxygen atoms. The atoms within each carbonate group are very strongly bonded to one another; the bonds between the carbonate groups and the calcium atoms are much weaker.

The figure above diagrams a prismatic crystal of calcite. The crystal structure is also indicated: blue circles represent calcium atoms; yellow circles indicate oxygen atoms; sets of three yellow circles represent the carbonate groups. Lines show the bonds between atoms.

The dashed line represents one cleavage plane that is perpendicular to the plane of the drawing. The cleavage plane falls along and cuts only the weakest bonds in the calcite structure. Other planes parallel to it are also potential cleavages. In addition, planes rotated by 120 or 240 degrees about the *c*-axis would also cut only weak bonds, and they are also cleavage directions.

Both the cleavage of a mineral and the faces shown by its crystals are important windows into the internal structure of the mineral.

Luminous Calcite and Aragonite: Fluorescence and Phosphorescence

Werner Lieber *of Heidelberg, Germany, explains a physical phenomenon for which calcite and aragonite are illuminating examples*

Banks and nowadays even gas stations employ ultraviolet radiation emitted by an ultraviolet lamp to test the authenticity of paper money. When an object is exposed to ultraviolet or UV radiation, previously invisible details might be enticed to luminesce. For example, US $20 bills show a bright green vertical stripe when exposed to ultraviolet; on a 20 Euro note, several parts of the design fluoresce green and tiny threads that are incorporated into the paper glow brightly in various colors.

Ultraviolet light is electromagnetic radiation whose wavelength falls between 100 and 400 nanometers, shorter than that of visible light and longer than X-rays. It is invisible to the human eye.

Luminescence is sometimes called *cold light*, as it is light that is generated from energy sources other than heat. The vast majority of substances do not luminesce under ultraviolet radiation, and the colorful glow of those that do, can often only be seen in the absence of ambient light.

When a suitable substance is exposed to ultraviolet radiation, its atoms or molecules absorb energy. The subject's electrons become excited as they are promoted to a higher energy level. Upon returning to their normal state, the electrons release some of the excess energy as visible light. The color of the visible light is related to the energy difference between the excited state and normal state.

Cambridge mathematician and physicist Sir George Gabriel Stokes (1819-1903) was the first person to recognize that the light emitted by a luminescent object always has a longer wavelength than the radiation used to stimulate that luminescence. The difference between the two wavelengths is known as the *Stokes shift*. In 1852, Stokes coined the term *fluorescence*, defined as the luminescence of a substance by electromagnetic radiation (in this case ultraviolet) in which light is emitted from a subject as it is exposed to radiation; light emission stops immediately when it is no longer exposed to radiation. *Fluorescence*, analogous to the term *opalescence*, was derived from the term *fluorite*, a mineral that from some localities has a particular propensity to fluoresce.

Some substances store a portion of the ultraviolet energy to which they are exposed. These hold the energy for intervals ranging from a few seconds to several hours, depending upon the substance. For

*Underground in the world's most famous locality for fluorescent minerals. This photo shows predominantly calcite (bright red) and willemite (yellow-green), but more than 60 fluorescent mineral species come from the **Sterling Hill Mine in Franklin, New Jersey.** Underground photography • field of view is roughly 4 m • photo archive Richard Hauck*

Calcite under Normal and Longwave Ultraviolet

Calcite twin from Dal'negorsk*; height 3 cm*

Calcite specimen from the Gotthard tunnel in Switzerland*; width 13 cm*

Calcite twin from Spain*; height 8 cm.   Paul Rustemeyer photos*

example, modern luminous watch dials, made of non-radioactive phosphorescent materials, and synthetic materials such as plastic Halloween decorations store ultraviolet energy for some time. When the source of radiation is removed and the substance continues to luminesce, the substance is said to *phosphoresce*. In general, the intensity of the luminescence diminishes quite quickly at first, then more and more slowly until all of the energy stored by the substance has been released.

Types of UV Excitation

A given substance will fluoresce or phosphoresce only when exposed to radiation of certain wavelengths. Xenon light sources and tunable lasers are used in research laboratories to produce radiation of specific wavelengths. Light sources employed by the hobbyist use less expensive mercury vapor lamps. Modern, low-pressure mercury vapor lamps emit shortwave ultraviolet intensely at 253.7 nanometers. By lining a conventional shortwave tube with a phosphor, longwave ultraviolet is produced from the same mercury vapor lamp. The phosphor absorbs the shortwave ultraviolet and emits longwave ultraviolet. Typically these lamps contain a filter to block visible light. Many minerals fluoresce under both wavelengths, though they do so with variable colors and intensities.

Type of Radiation	Wavelength Emitted (in nm)
Shortwave ultraviolet	sharp peak at 254
Longwave ultraviolet	broad peak at 366
Visible light	spectrum from 400 to 700

Table 1: Type of radiation and wavelength it emits

Because fluorescence tends to be fairly weak when compared to visible light, it is best observed in a dark room or inside a test box. It is advisable to allow time for one's eyes to adapt to the darkness, and it is important to wear UV-absorbing glasses to prevent eye damage from exposure to shortwave ultraviolet. Longwave ultraviolet causes the interocular fluid of the eye to fluoresce, interfering with one´s ability to see true colors of fluorescence.

The Role of Impurities

Calcite and aragonite are among those minerals that both fluoresce and phosphoresce. Fluorescence is not, however, a defining characteristic of calcium carbonate, as not all specimens fluoresce. White, gray or otherwise colored crystals as well as massive calcite such as that found in limestone and marble are more likely than colorless crystals to fluoresce. Numerous studies confirm that chemical impurities (e.g. the substitution of Mn^{2+} for Ca^{2+}) are responsible for either activating or quenching fluorescence. Divalent iron (Fe^{2+}) is a notable quencher of fluorescence, however trivalent iron (Fe^{3+}) is an activator in such minerals as feldspars and scapolite, where it substitutes for Al^{3+} and causes red fluorescence in these minerals. Activators are often the transition elements of the periodic table. Optical analyses of the wavelength of the absorbed light, such as those performed with a spectrograph, can help to identify which atoms cause fluorescence.

Sometimes activators are an intrinsic part of the chemical formula of a species, in which case samples of that species always fluoresce unless a quencher is present. Scheelite ($CaWO_4$) is one such example: the tungstate radical (WO_4) is the activator and is always present. In other species such as calcite and aragonite, impurity activators may or may not be present, and samples in turn may or may not be fluorescent. When impurities are present, they commonly occur in amounts too small to be included in the chemical formula of the species. For many substances, there is a long list of elements, ions or molecules that are known or suspected activators of fluorescence.

Examples of Fluorescence in Calcite

The intense red fluorescence of calcite from the former zinc mines at Franklin and Sterling Hill, New Jersey is truly spectacular. Here, fluorescence is activated by small amounts of manganese, ranging from 0.2 to several weight percent, that contaminate the calcite lattice. Calcite samples containing around 1.8 weight percent manganese (equivalent to 3.6 percent manganese carbonate) luminesce with particular intensity. Calcite samples whose manganese content is higher than 1.8 weight percent exhibit a diminished fluorescence that disappears completely when the content of manganese carbonate exceeds 6 weight percent (Köhler & Leithmeier, 1933; Mutschler, 1954). The in situ exposure of fluorescent willemite and calcite in the Rainbow Room at the Sterling Hill mine, part of the Sterling Hill Mining Museum, is a beautiful showplace of fluorescence.

Because its ionic radius is similar to that of calcium, divalent manganese is easily accepted into the

To Excite Ultraviolet Glow

Mineral	*Activator*	*Color of Fluorescence*
Aragonite	unknown	white, cream, pink (sw.lw)
Calcite	Mn^{2+}	orange-red (sw)
Calcite	Eu^{2+}	blue (sw), pink (lw)
Calcite	unknown	yellow, white (sw,lw)
sw=shortwave; lw=longwave (Modreski, 1999)		

Table 2: Known or suspected activators for calcite and aragonite (after Modreski, 1999)

calcite lattice. Samples of manganese-containing calcite from many other localities fluoresce in a diversity of reddish colors and intensities. Mixed colors can also occur due to the presence of other activators.

Calcite from Andreasberg, Germany glows a soft red under longwave ultraviolet radiation, as do many of the crystals from Elmwood, Tennessee. White calcite crystals in the amethyst geodes from Irai in Rio Grande do Sul, Brazil fluoresce a strong orange-red, as do some colorless and white, twinned crystals from Dal'negorsk in Primorye Kray, Russia. White or light gray crystals from Chenzhou, Hunan, China glow a similar red. Calcite from Iberg in the Harz Mountains in Germany glows green to yellow-green under shortwave UV, but white to yellow-white under longwave UV; these specimens also exhibit strong white-yellow phosphorescence. Colorless calcite from Tsumeb, Namibia also glows weakly red under longwave UV but is significantly brighter under shortwave. Some calcite samples from the famous mercury deposits at Terlingua, Texas fluoresce blue under shortwave, followed by intense phosphorescence. They are pink under longwave ultraviolet light.

Examples of Fluorescence in Aragonite

Aragonite from many localities is also interestingly fluorescent. The beautiful crystals from Horenice near Bílina, in the Severoceske region of the Czech Republic, glow intensely green under shortwave because of microscopically thin coatings of uranyl-activated hyalite opal. The aragonite glows intensely white under longwave ultraviolet radiation. The well known aragonite trillings from Agrigento in Sicily, Italy are also wonderfully fluorescent: they glow hot-pink under longwave and white under shortwave, phosphorescing intensely green when the lamp is extinguished.

Even a relatively innocuous camera flash is well known to produce plenty of ultraviolet radiation that can invoke this splendid phosphorescence. The green-glowing specimen on page 39 was photographed with daylight film immediately after being exposed to multiple flashes with the camera's shutter closed. Depending upon the intensity of the flash, phosphorescence generally lasts about five seconds, with steady reduction in luminosity.

The El Dorado for collectors of fluorescent minerals: the Sterling Hill Mine near Franklin, New Jersey, USA

The famous aragonite (var. *tarnowitzite*) trillings from Tsumeb, Namibia exhibit beautiful fluorescence and phosphorescence: under shortwave ultraviolet radiation, they glow light yellow to green, while under longwave ultraviolet they are strongly yellow.

An Interesting Field for Hobbyists

The examples described here represent only a small sample of the localities that produce fluorescent calcite and aragonite. The colors described are not constant; individual crystals fluoresce differently even from the same locality. While it can only rarely be used for reliable confirmation of locality information or to test a species' identity, the fluorescence of minerals is an interesteing phenomenon and an important aspect of the hobby.

Calcite and Aragonite Under Ultraviolet

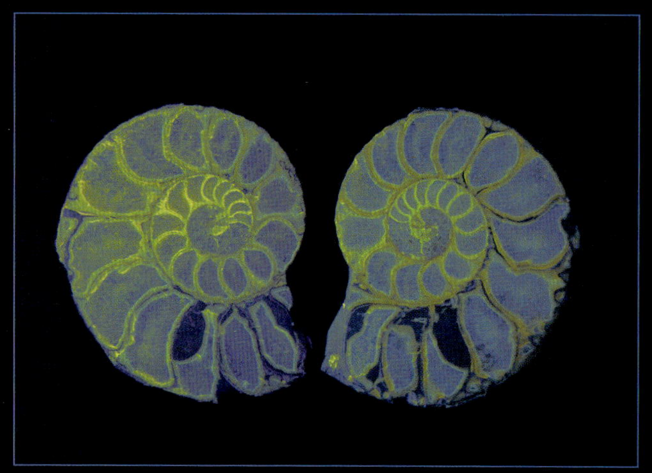

Sedimentary calcite under ultraviolet light *• Left: Ammonite; (4 cm); right detail of a septaria (7 cm)*
Franziska von Kracht collection; Maximilian Glas photo

"Tarnowitzite"*, the lead variety of aragonite from Tarnowitz, Poland; field of view 5 cm*
Bavarian state collection; Maximilian Glas photo

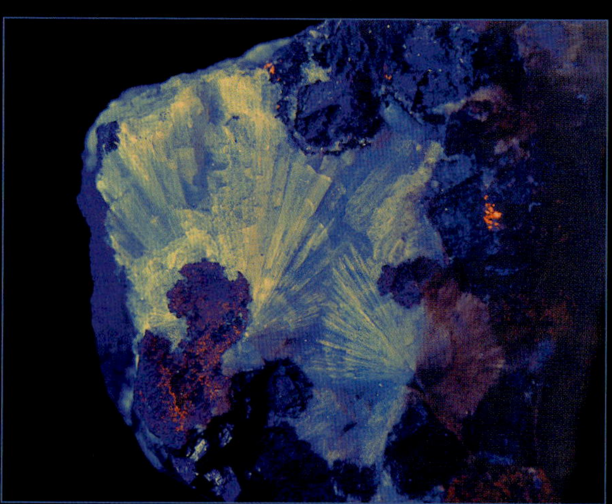

"Tarnowitzite" *from Tsumeb, width 6 cm*
Bavarian state collection; Maximilian Glas photo

Phosphorescent Aragonite

Upper:

Aragonite crystals from
Agrigento, Sicily, Italy
under ultraviolet light
field of view 12 cm
Werner Lieber photo

Left:

Calcite and aragonite from
Finale Ligure, Savona, Italy
width of specimen 7 cm

The upper picture was taken under normal lighting conditions, the lower picture shows the glow (phosphorescence) after exposure to ultraviolet radiation
Paul Rustemeyer photo

A Lucky Break For Polarization:

Professor and researcher at the University of Idaho in Moscow, Idaho (USA) ***Mickey E. Gunter*** *brings the optical properties of calcite into clear view.*

Were it not for the fact that the cleavage direction of calcite is neither parallel nor perpendicular to its optic axis (used herein as synonymous to the *c*-axis), it might have been centuries before scientists discovered the polarization properties of light. In addition, if calcite did not have such a large difference in refractive index values, double refraction may not have been observed. In a way then, it was a lucky break for polarization and for science that common calcite has some of the properties that it does.

When in 1669 Erasmus Bartholinus observed an image through a transparent cleavage rhombohedron of calcite, he saw that the image was doubled; this doubling led him to coin the term ***double refraction***. Both of the calcite rhombohedra in Figure 1 illustrate this phenomenon.

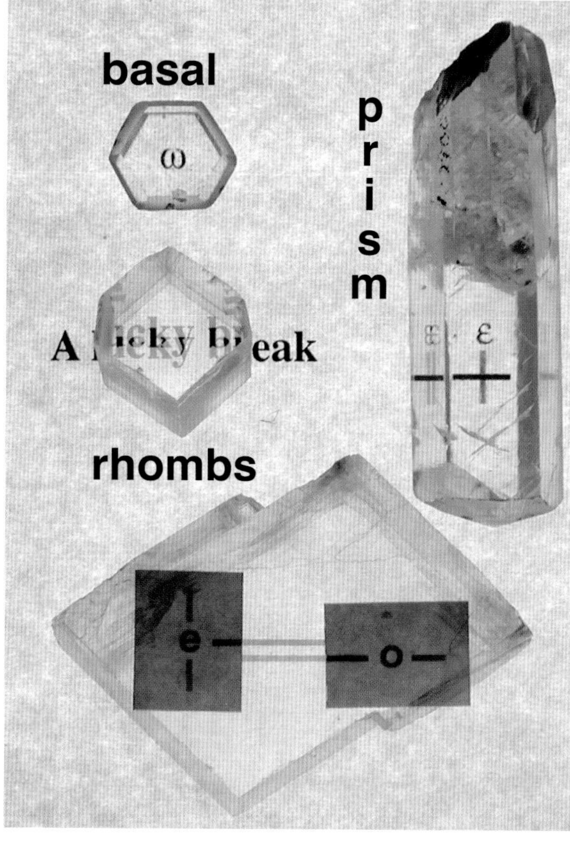

Figure 1: Light Propagation Through Three Habits of Calcite *Light travels down the c-axis of the pinacoid and remains unpolarized. The c-axis of the prism is parallel to the page. Light passing through the crystal is forced to vibrate along either the ω or ε direction*

Photo by Mickey Gunter, samples were provided by Harvard Mineralogical Museum and the Natural History Museum of Los Angeles

Interestingly, Figure 1 contains three habits of calcite, but only the cleavage rhombohedra shows the effect of double refraction first observed by Bartholinus.

Bartholinus' discovery was the beginning of our understanding of the polarization properties of *anisotropic* minerals, minerals through which the speed of light varies as a function of direction. The speed of light is the same regardless of direction in an *isotropic* material. Ironically many if not most professional and amateur mineralogists, as well as most physicists, believe that double refraction occurs anytime an anisotropic mineral is not viewed along its optic axis. This, however, is not necessarily the case.

Bartholinus' observation of double refraction conflicted with Willebrod Snellius' widely accepted 1621 law (Kile, 2003) that predicted how light would be refracted (i.e., bent) as it traveled between materials with different refractive indices. *Snell's Law* is written

$$n_i \sin \theta_i = n_r \sin \theta_r \qquad \text{Equation 1}$$

where n_i is the refractive index of the incident ray (i.e., the refractive index of the medium through which the ray travels before striking a material with a different refractive index), θ_i is the angle of incidence as measured from a normal (i.e., a perpendicular line between the surfaces of the two materials with different refractive indices), n_r is the refractive index of the refracting media and θ_r is the angle of incidence of the refracted ray.

The refractive index of a material is the speed of light in a vacuum divided by the speed of light in the material. This ratio is always greater than 1.0 because as the photons of a light beam interact with the electrons of a material, the photons are slowed. The greater the interaction between the light beam and the material, the higher the refractive index of the material.

Figure 2 illustrates the refraction of ray 1 as it travels from air into a piece of glass at an angle of 45 degrees. The precise angle of refraction can be calculated using Equation 1 (assuming one knows the refractive index of the glass). When a light ray travels from a material of lower to one of higher refractive index, it is refracted toward the normal of the latter surface. When light travels from a medium with a high refractive index to a medium with a low refractive index, the ray is refracted away from the normal; however, for a ray at normal incidence (ray 2 in Figure 2), no refraction occurs; light travels straight through the material undeviated, though it is slowed.

According to Snell's Law, only one image should be produced when light is normally incident on a transparent material. When light enters normal to a calcite rhombohedron face, however, two images

the Optical Properties of Calcite

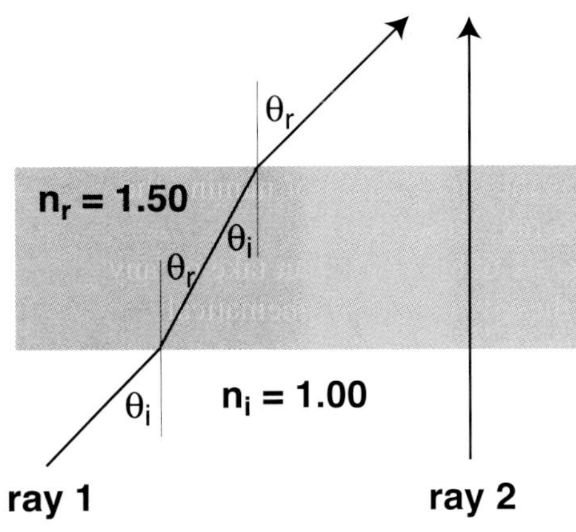

Figure 2

are produced, clearly a violation of Snell's Law. As will be discussed later, light entering either the pinacoid or the prism faces of calcite does not produce two images.

Bartholinus based his nomenclature for the rays of light on this perceived discrepancy with Snell's Law created when light entered normal to the rhombohedron. He called the ray that obeyed Snell's law the *ordinary ray*, abbreviated *o-ray,* and referred to the ray that did not obey Snell's Law as the *extraordinary* or *e-ray*. These terms came to be used for the refractive index values for the entire uniaxial class of minerals, ω and ε. As we will see, however, both ω and ε are actually o-rays, as they do conform to Snell's Law.

Polarization and Calcite

In the early nineteenth century, explaining double refraction in calcite became a major goal for the scientific community. In 1807, the French Academy of Sciences offered a prize for anyone who could explain the phenomenon. Étienne Louis Malus sought the prize (Kristjansson, 2002) but while studying calcite, discovered instead polarization by reflection of light. Malus coined the term *polarization*, but it appears to be Augustin Jean Fresnel and François Jean Dominique Arago (Bloss, 1999) who realized that calcite polarized light. In fact, they showed that the o-ray and e-ray were polarized at right angles to one another. This effect is seen in the larger rhombohedron in Figure 1 by placing Polaroid sheets on it. These Polaroid sheets only allow light polarized along their long dimension to pass. The sheet on the left allows the e-ray to pass; the one on the right allows the o-ray. One of the two images in the center of the rhombohedron is eliminated by the sheet polarizer.

Three orientations are of interest for an unpolarized light beam as it enters a calcite crystal: 1) normal to a basal section (i.e., parallel to the optic axis), 2) normal to a prism face (i.e., normal to the optic axis) and 3) normal to any other plane, such as the face of a scalenohedron or a rhombohedron (fortunately the most common case). All three of these orientations of light are shown entering natural crystals in Figure 1. Sketches of each light path are shown in Figures 3 and 4 on page 42. The situation illustrated by the rhombohedral face is the general case, in which light travels neither parallel nor perpendicular to the optic axis.

When unpolarized light enters normal to the rhombohedral face, it is split into two mutually-perpendicularly polarized rays, the o-ray and the e-ray (Figure 3). The o-ray proceeds undeviated through the sample: it has a polarization direction perpendicular to the optic axis. The e-ray vibrates in the plane of the optic axis, parallel to the rhombohedral face. As the rays emerge from the crystal, the o-ray and the e-ray are thus offset, producing two images that are polarized perpendicular to one another. Explanation of double refraction led to the discovery that calcite, and all anisotropic minerals, can polarize light.

Light traveling perpendicular to a basal section of calcite is traveling parallel to the optic axis of the crystal. By definition, the **circular section** of a uniaxial mineral runs perpendicular to its optic axis. **The circular section of a uniaxial mineral has the same refractive index in all directions;** thus, it behaves like an isotropic mineral. Entering normal to a basal section, unpolarized light is not polarized as it passes through the crystal.

A similar but slightly different phenomenon occurs when light enters a crystal normal to a prism face (Figures 1 and 4). Again, double refraction does not occur, but in this case the incident unpolarized light is polarized as it passes through the crystal, and two images are produced. One ray travels through the crystal faster than the other; consequently a dot below the crystal is seen as two dots that are aligned one on top of the another.

Were the pinacoid or the prism the only forms of calcite, double refraction may not have been discovered, at least in calcite, until someone had cut the mineral at an angle random to the *c*-axis.

Figure 2: *Ray tracing through a block of glass showing the angular relations based on Snell's Law. Ray 1 is incident on the block of glass at the angle θ_i and is refracted to the angle θ_r after passing into the glass. Ray 2 is normally incident on the glass, and undergoes no refraction.*

Drawing by the author

A New Look At An Old Rule: Snell's Law Redefined

In Figure 3, the side view of calcite has an extra ray labeled WN (wave normal). The wave normal is the direction perpendicular to the vibration direction of polarized light. Notice the wave normal for the e-ray passes undeviated through the calcite rhombohedron, thus obeying Snell's Law. If, therefore, Snell's Law had been defined based on wave normal and not on ray paths, it could have been applied to any direction in an anisotropic mineral. Snell's Law was derived for light passing through isotropic media, and calcite cleavages appeared to violate this law.

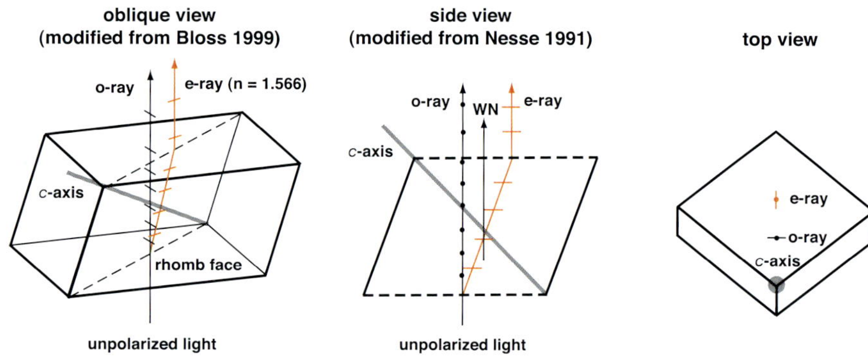

Figure 3 *(above): Three views of a calcite rhombohedron and the associated light paths after passage of a normally incident, unpolarized light beam. The oblique view shows the incident ray being split into two mutually perpendicular polarized rays. The polarization direction of each beam is indicated by dashes along the ray path. Both rays are contained in the dashed plane cutting the center of the rhombohedron that contains the optic axis. The side view shows the ray paths and associated vibration directions. The o-ray vibration direction is perpendicular to the page. The top view shows the e-ray and o-ray projected on the top surface of the rhomb. If the rhombohedron is rotated on the page, the o-ray stays stationary while the e-ray orbits around it.*

According to the current notation the two principal refractive index values for a uniaxial mineral are ε and ω, with any intermediate value termed ε'. This terminology was the direct result of naming the rays e-ray and o-ray. For calcite, ε = 1.486 and ω =1.658; thus, ε' can take on any value between these two. The mathematical relationship between these values is:

$$\varepsilon' = (\omega^2 \cos^2\theta + \varepsilon^2 \sin^2\theta)^{-1/2}$$

where θ is the angle between the optic axis and ε'. For the rhombohedron face, the ε' direction makes an angle of 44.6 degrees to the optic axis; thus, ε' = 1.566.

One of the problems with the evolution of the nomenclature for uniaxial optical values is that often the *e-ray* is used synonymously with the ε refractive index value. The labeling in Figure 4 of the side view of the prism face may appear to be in error in that both ω and ε values are labeled as *o-rays*. The figure labeling is correct, however, in spite of the fact that ε is often mistakenly thought to be an extraordinary ray. When a light ray is perpendicular to the optic axis in a uniaxial mineral, both rays behave as ordinary rays in that the paths of both obey Snell's Law.

All drawings are by Mickey Gunter

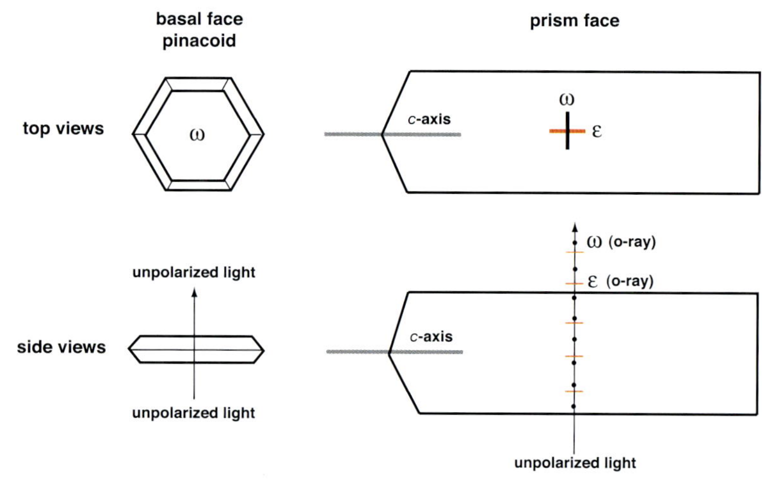

Figure 4 *(left): Top and side views of pinacoid and prism habits of calcite. In the pinacoid, the c-axis (optic axis) is perpendicular to the page and unpolarized light passes through the sample and remains unpolarized. For the prism the c-axis is parallel to the page and a normally incident unpolarized light beam would be broken into two mutually perpendicular polarized components vibrating along the ε and ω values. Note that in this orientation both rays would behave as o-rays.*

Literature for further reading:

The author recommends FD Bloss' 1999 book *Optical Crystallography* as well as WD Nesse's 1991 *An Introduction to Optical Mineralogy* to anyone who like to learn more about this subject.

Double refraction occurs only when the incident light enters the crystal in an orientation that is neither parallel nor perpendicular to its optical axes. This is true not only for calcite but for every uniaxial mineral.

Calcite in Polarizers

As noted above, when a ray moves from a material of high refractive index to one of low refractive index, it is refracted away from the normal. There is, thus, some angle of incidence at which the angle of refraction would equal 90 degrees. When the angle of refraction is equal to 90 degrees, none of the ray is transmitted through the medium of lower refractive index; it is entirely reflected back into the medium with the higher refractive index value.

Scottish physicist William Nicol took advantage of this and the fact that calcite yields two polarized waves of greatly different refractive index and in 1829, produced the first polarizing prism. He built the Nicol prism by cutting a calcite rhombohedron diagonally in half then gluing the two pieces back together in the same orientation.

As light enters the Nicol prism (Figure 7), unpolarized light is, as usual, broken into two mutually perpendicular polarized waves with different refractive indices: $\omega = 1.658$ and $\varepsilon' = 1.516$. When the ω component strikes the balsam interface glue ($n = 1.537$), it undergoes total internal reflection and does not cross the balsam - calcite boundary; thus, only the e-ray travels through the second portion of the rhombohedron and emerges as a single polarized ray. By 1860, Nicol prisms were incorporated into light microscopes, creating the first true polarized-light microscopes. They were used to study minerals and rocks (Kile, 2003).

A polarized-light microscope provides information beyond what is provided by an ordinary light microscope. It reveals invaluable information about the structure and composition of minerals. Through it, one can distinguish, for example, between isotropic and anisotropic materials.

Figure 5: *A Nicol prism on the left and two Glan-Thompson polarizers removed from a polarizing light microscope on the right.*
Photo Mickey Gunter

Polariscope, circa 1890
Precursor of the modern petrographic microscope, this polariscope, engraved by R. Fuess, Berlin-Steglitz, employs a large Nicol prism in the stage as a polarizer. The small Nicol prism in the ocular position is the analyzer (height 37 cm).
Photo Kubath and Medenbach

Figure 6: Through the Polarizing Microscope
This calcite rhombohedron thin section from the Munich (Germany) Gravel Flats exhibits mechanical twinning; magnification 45x, crossed-Nicols
Günther Neumeier photo

Figure 7: Ray paths for the Nicol (left) and Glan-type (right) polarizers; each takes advantage of total internal reflection of ω to produce a polarized beam. The Glan-Foucault polarizer has air between the cut portions of calcite, the Glan-Thompson polarizer is cemented together with a higher refractive index material; thus, the acute angle of the calcite prism must be different.

Sketches by Mickey Gunter

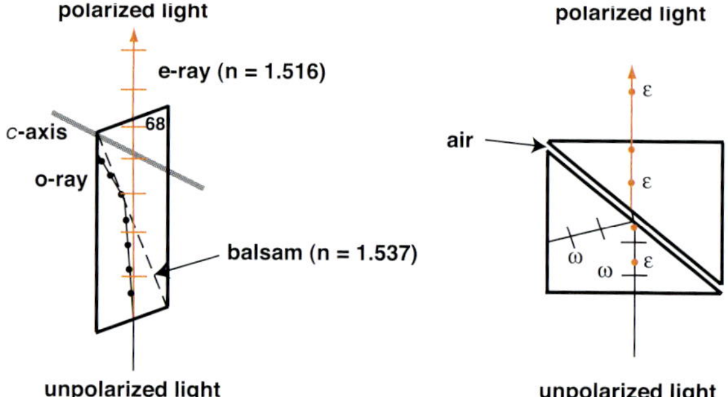

Figure 7

The Nicol prism was replaced in the late 1800s by the Glan-Thompson prism, which was among other things easier to manufacture. The Glan-Thompson prism was also based on total internal reflection of the ω ray. It was made of blocks of calcite that were cut in such a way as to transmit the light perpendicular to the optic axis, similar to the way that light is transmitted through the calcite prism face in Figure 4. Calcite for the Glan-Thompson prism was cut diagonally parallel to the *c*-axis; the ω ray is reflected at this boundary. In Figure 7, air is between the two pieces, but different types of glue can be used, which in turn would require the diagonal angles to differ.

Polaroid sheets, developed in 1935 by Edwin H. Land (Kile, 2003), replaced calcite-based polarizers in most applications. Although Nicol prisms have not been widely used for nearly a century, microscopists often use the term *crossed-Nicols* to denote a sample viewed in cross-polarized light.

Calsight

The first interference figure for a mineral was observed by Sir David Brewster in 1818 when he viewed a piece of calcite that was sandwiched between crossed polarizers and was illuminated in convergent light. This effect is easily reproduced by placing a 2 to 3 mm thick basal section of

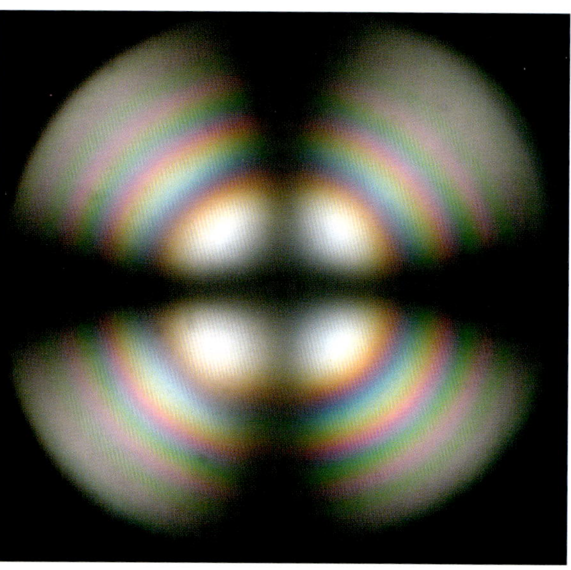

***Figure 8:** Optics Axis Figure*

calcite, which is at least 5 mm in diameter, between two sheets of Polaroid whose vibration directions are at right angles to one another.

From a distance of 1 meter, the calcite-polarizer sandwich appears dark, but moving the combination to within one inch of the eye, an interference pattern similar to the one shown in Figure 8 appears. Interference figures aided scientists in identifying and characterizing minerals. Using interference figures, mineralogists derived three optical classes: isotropic, uniaxial and biaxial. These were later related to the crystal structure of the minerals.

During the early stage of World War II, the US military asked Land to develop a new type of optical sight that could be used for artillery (Orrell, 1993). The military wanted a single-element sight to replace the old system that required a front and rear sight. Among other advantages, the single-element sight would not create parallax. For his sight, Land used the same set-up that was used to create interference figures: a basal section of calcite

Timeline (with associated direct and indirect references) of calcite's role in significant developments of polarization, (modified from Kile, 2003)

1621	1668	1669	1808	1811	1818	1829	1860
Willebrod Snellius develops Snell's Law, a mathematical expression that relates the angles of refraction of a light ray to the refractive indices of a media.	Earliest confirmed finds of *Iceland Spar* from Helgustadir, Iceland	Bartholinius first reported the phenomenon of *double refraction* that he observed in *Iceland Spar*	Malus discovered and named polarization while studying double refraction in calcite	Fresnel and Arago observed that the two images refracted by calcite were polarized	Brewster discovered that optic axis interference figures formed when a calcite plate was illuminated with convergent light	Nicol invented a polarizar (the Nicol prism) consisting of two pieces of calcite oriented to produce a single beam of polarized light	First polarizing microscope to use the Nicol prism
(Kile, 2003)	(Kristjansson, 2002)	(Bartholinus, 1669; Kristjansson, 2002)	(Malus, 1808; Kile, 2003)	(Bloss, 1999)	(Brewster, 1818; Kristjansson, 2002)	(Nicol, 1829; Kile, 2003)	(Kile, 2003)

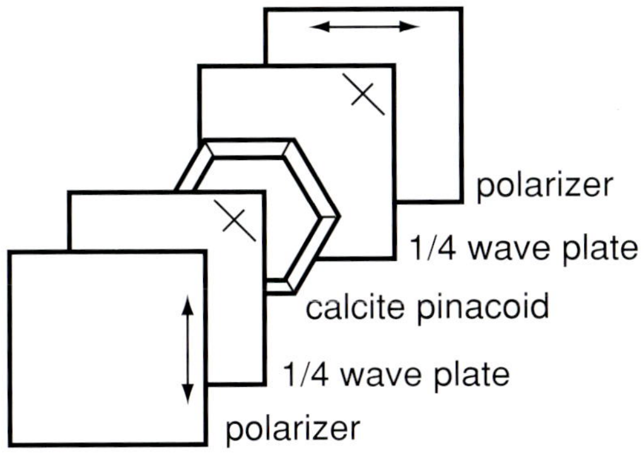

***Figure 9:** Elements of An Optical Ring Sight*

Development of the Optical Ring Sight (ORS)

Figure 8 is the optic axis figure of calcite that forms when a cone of converging light rays strikes a pinacoidal face of calcite that is sandwiched between two cross-polarizers. The isogyres (black cross) of the interference figures can be removed, yielding the optical ring sight (Figure 10), by placing two oriented 1/4-wave retarders in front of and behind the calcite as is Figure 9.

Photos and sketches by Mickey Gunter

sandwiched between cross-polars. He wanted, however, to remove the cross. Land therefore added plates of anisotropic material for which the slow ray lagged 1/4 wave behind the fast ray. He oriented these *1/4 wave retarders* so that their slow vibration directions were parallel to one another and 45 degrees from the vibration directions of the polars (Figures 9 and 10).

Having developed a sight, Land needed a US source of optically clear calcite in order to manufacture the device. Mineralogists at Harvard assisted in a search and the Palm Wash calcite deposit was located near Salton City in San Diego County, California (Orrell, 1993). One major advantage to the calcite found at this deposit was that it formed with a habit dominated by the basal pinacoid, making sample preparation much easier. Land's ring sights were used during World War II and are still in limited use today.

1881	1935	1945
S.P. Thompson invented the Glan-Thompson polarizer	E.H. Land invented the first sheet Polaroid	First commercially produced polarized-light microscope to use sheet Polaroid
(Kile, 2003)	(Kile, 2003)	(Kile, 2003)

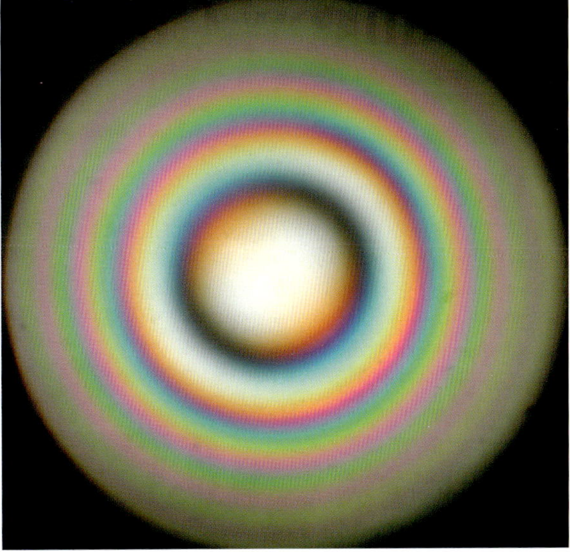

***Figure 10:** Optics Axis Figure With Isogyres Removed*

An exhibit at the Smithsonian Institution in Washington D.C., shows a modern, commercially available optical ring sight and discusses the sight's past and present use.

The photo is by Dane Penland and was provided by Jeff Post of the Smithsonain Institution in Washington, D.C.

Acknowledgements

The author would like to thank Carl Francis and Anthony Kampf for providing calcite samples, Jeff Post for the photo of the optical ring sight display, and all those who responded to his questions about ring sights that were posted to the Mineralogical Society of America's list server, but especially George Rossman, Carl Francis, Mac Ross and David Vanko. He also extends his thanks to Bill Mabbutt of Gem State Crystal in Moscow, Idaho for aiding in sample preparation and Suzanne Aaron for serving as his technical editor for the past 20 plus years.

Rainbow Calcite: A Colorful Effect of

Missoula, Montana lapidary and gemstone dealer ***Michael Gray*** *with his take on cutting kaleidoscope calcite gems*

One does not ordinarily think of calcite as a faceted gemstone; yet, calcite is one of the most spectacular collectors' gems. While passing the display cases at a gem and mineral show, you may have been struck by the spectrum of colors emanating from certain faceted calcites. Instead of small points of primary colors such as those seen in diamonds, imagine large flashes of pinks, greens, blues and yellows, what I call *kaleidoscopic colors*, dancing around an otherwise colorless stone. You may have wondered briefly why some stones show such a play of color. "Is it the lighting, the cut or the raw material?" The answer is, "A little of everything."

***Upper**: "The Doris," a 1017- carat calcite gemstone cut by Art Grant, is on exhibit at the A.E Seaman Mineral Museum. The rough for this stone came from Balmat, NY.*

***Middle and lower :** Two views of a 246.5-carat stone cut by the author. All three photos are by Jeff Scovil.*

A Strategically Oriented Twin Plane

What makes these mostly colorless stones so colorful when cut is the presence of a twin plane precisely oriented within the gem. Only cut calcite with one or more oriented twin planes will exhibit this rainbow effect. Of course, a good cut (flat facets and no windowing) and polish are key elements as well. The rainbow colors are seen almost exclusively in colorless calcite, but there are a few exceptions, as the occasional light yellow stone gives off a spectacular display.

My dad Elvis *Buzz* Gray and I chanced upon this phenomenon in the early 1970s while we were cutting a number of pieces of colorless calcite rough from Mexico. Through trial and error, we discovered a range of orientations of the twin plane(s) that would maximize the flashes of color. We found that to achieve the optimum color display, the twin plane should be placed at an angle that is between 20 and 42 degrees from the table of the stone. In addition, the twin plane should pass through the lowest row of the stone's facets, as other orientations tend to result in less color play. The kaleidoscope effect is magnified when there are multiple twin planes present within the crystal, especially when they are not parallel to one another. In stones with too many twin planes, say five or more, the colors are still present, but the finished stone looks hazy.

Five or More Twin Planes?

Twins that develop during the growth of the crystal always have a single twin plane; however, twinning can also be produced by pressure that is applied to a crystal after it forms, for example by distortion of the

surrounding rock by metamorphism, or even by the shock of the blast that reveals the crystal. Calcite responds to the pressure by slipping

Reflection, Refraction and Twinning

along specific planes in the crystal, changing shape slightly to reduce the pressure and leaving thin slices that are twinned relative to the rest of the crystal. Calcite crystals with this form of *mechanical twinning* can have multiple twin planes at different angles, though all of the planes are parallel to one of the faces of the form $\{01\bar{1}2\}$. Either growth or mechanical twin planes can produce the kaleidoscopic play of colors seen in these cut stones; however, growth twins on {0001} are not optically active in the same way.

Light Rays Split Again and Again

In my opinion, the best article covering this effect was written by Cornelius Hurlbut, Jr. and Carl Francis, both of Harvard University. The article was published in the Winter 1984 edition of *Gems & Gemology*. It describes a particularly spectacular 1,156-carat stone that was cut by Art Grant. This stone is a growth twin on $\{01\bar{1}2\}$ from the Faraday Mine in Hastings County, Ontario, Canada. It has a single twin plane oriented at about 67 degrees to the table of the stone.

In the article, the authors describe the features that cause of the complex play of color, explaining that light enters the calcite through the table and is split into two rays (the familiar double refraction of calcite, see page 40). These rays travel through the stone in slightly different directions. When the rays of light encounter the twin plane, each is split again into two more rays. The four light rays each take a slightly different path through the crystal. Each of these reflects back from the facets and is reflected again, this time upward toward the table where the split rays of light emerge to present the viewer with an image of the facet from which each was reflected. Before reaching the table, however, the rays pass once more through the twin plane and are split yet again.

In the end, there are eight rays of light, each traveling through the stone on a different path. Calcite's optics cause each light ray to carry not only a different image but also a different color. Some of these images partially overlap and create still different colors. The colors presented depend upon the angle at which the stone is viewed. As the stone is rotated slightly, the colors change: truly a kaleidoscope!

Colorless Rough Cuts the Most Colorful Gemstones

I have experimented with other gem materials that exhibit twinning, such as *sphene* (titanite) and *kunzite* (spodumene). Cut with a properly oriented twin plane, these species also transmit kaleidoscopic colors, though generally to a lesser degree than does colorless calcite, as the rainbow effect in other materials is masked by the body color of the stone.

The largest, clean calcite cut to date that exhibits the rainbow effect weighs 4,630.75 carats. The rough came from Balmat, New York. I cut the stone in the early 1980s. It is now in Michael Scott 's collection, which is on display at the Bowers Museum in Santa Ana, California

Upper:
The rough for this 37.32-carat, cut, twinned calcite came from Yakutia, Russia. Photo and collection Terry Huizing

Middle and lower:
A view of the top and the bottom of the same 523.08-carat stone fashioned by author Michael Gray from a piece of Brazilian rough. Photos Jeff Scovil

The author wishes to extend special thanks to editor Pete Richards for the invaluable contribution that he made to this text

China: a Promising New Entry in

Coal geologist and Hunan native ***Guanghua Liu****, currently living in Tübingen, Germany, with a first-hand look at China's most important calcite localitites.*

Over the last decade, a wide variety of Chinese calcite specimens have turned up on the international mineral market. The language barrier between China and the west as well as some dealers' reluctance to accurately label specimens have produced widespread confusion regarding Chinese localities among collectors.

Traditionally, the Chinese do not collect mineral specimens, though they have a great passion for jade, for unusually shaped and colored stones and for carvings. With China's opening to the west in the 1980s have come new opportunities for collectors in search of Chinese specimens. This in turn has led to the development of specimen mining in China and to the emergence of trading centers in a number of areas. These are mainly concentrated in southern China, in the Hunan Province and Guangxi Zhuang Autonomous Region. Although calcite occurs in many of China's provinces, this review will focus on the localities most heavily represented in today's specimen market.

White Roses *Found in the Shangbao pyrite mine, clusters of pure white "roses," like Francis Benjamin's 7 cm wide specimen pictured above, had collectors in the mid-1990s clamoring for more. Photo Jeff Scovil*

Right: The entrance to the Shimen realgar mine as photographed in 2003 by Rock Currier

Hunan: New Finds in Ancient Mines

Hunan is a mountainous inland province in southern China, with 210,000 square kilometers thinly populated by a mere 65 million people, mostly of the Han ethnic group. Other minority groups such as Tujia, Miao, Dong, Yao and Hui make up 8 percent of the provincial population. The province holds national importance as the birthplace of a number of China's modern day leaders including Chairman Mao Zedong. It is also a major rice producer, but to collectors the province is most well known for mineral specimens.

Soon after China opened its doors to the western world, light yellow, translucent, lustrous calcite clusters that were associated with realgar began to turn up at the shows in Munich and Tucson. These were the first Chinese calcite specimens to draw the attention of collectors. They came from the **Shimen (Stone Door) realgar mine** (formerly the Jiepaiyu realgar mine), situated 33 kilometers northwest of the town of Shimen and about 95 kilometers from the city of Changde.

One of the largest realgar mines in the world, the Shimen mine has been operating for more than 1500 years. Realgar and orpiment have long played an important role in Chinese medicine, and for generations this mine has been providing realgar, orpiment and arsenic products to all of China as well as to many other countries.

The realgar deposits occur in Lower Ordovician

the Mineral Specimen Market

limestone and Upper Cambrian dolomite formations and formed as the result of low-temperature hydrothermal activity. The orebodies developed in cavities and veins that are concentrated along tectonic fracture zones of the carbonate rocks. Mineralization is dominated by realgar, orpiment, calcite, dolomite and pyrite. These are occasionally associated with stibnite and quartz crystals.

A considerable number of superb specimens of realgar, orpiment, calcite and associated species were found in the 1990s, but a few years ago the realgar veins were exhausted and the mine closed. The former miners, however, continue to search for specimens. In November 2002, they discovered pockets with lustrous clusters of calcite crystals, and some specimens from that find were available at the Tucson show in February of 2003. Though no significant realgar crystals were found in these pockets, some calcite crystals had inclusions of realgar.

In the winter of 1997/1998, large, bicolor calcite clusters were found at the **Leiping mine**, in Guiyang County, Hunan. This mine is run by the local government and is located in the northern part of the county, 40 kilometers southwest of the city of Leiyang in southern Hunan. The mining region is peppered with non-ferric mineral deposits and mines including the Dashunlong copper-dominated, multiple metallic deposits in the north and the Baoshan copper-lead-zinc deposits in the south.

These deposits formed mainly by hydrothermal activity related to Paleozoic and Mesozoic granitic intrusions. The ore-bearing units in the Leiping mine are generally hosted in Devonian carbonates and granites. The deposits are rich in tungsten, tin, copper, lead and zinc minerals and have been interpreted as both hydrothermal and contact metamorphic.

Calcite is a common associated mineral from the Leiping mine and is mainly composed of steep rhombohedra that present a triangular cross section. Crystals range from a few centimeters to over 40 centimeters in length. The crystal clusters are very attractive and show pink to red color on one crystal face and are white or colorless on the other two faces. The red face is the result of hematite inclusions and may indicate the direction of the source, during crystal formation, of the iron-bearing fluid. The mine produced more than two tons of good quality calcite specimens. Most of this find was on the market in Tucson in 1997. There have been no further finds of these clusters of calcite.

More recently, various other habits of calcite have been found at the Leiping mine and in the surrounding areas. The most important discovery may have been the reddish to clear twins and clusters that were found in the summer of 1999 (cover photo). The twins are contact and penetration twins on $\{10\bar{1}1\}$, a relatively rare twin law for calcite (see page 11). These superb calcite twins and the high prices that they commanded encouraged the miners and farmers to concentrate their efforts on further prospecting.

Fire and Ice

In the fall of 1997, German dealer/collector Jürgen Tron acquired this 24.5 cm high red and white calcite cluster from the Leiping mine. The red faces are the result of hematite inclusions. Jeff Scovil photo

In the months that followed, large, tabular and prismatic hexagonal crystals, intersecting twins and attractive clusters were found in the nearby **Baisha quarry**. These specimens ranged from colorless and pale white to yellow, red and brown. Thousands of columnar, intersecting twin crystals as well as lustrous, deep red, brown and black calcite clusters were found at the **Daba Tin mine** in Leiyuan, just 15 kilometers away from the Leiping mine. Green short pyramidal crystals and clusters of calcite as well as green needles of aragonite have also been found in the nearby copper-tin-zinc **Dashunlong mine**. Some of these were associated with massive malachite. Specimens from this mine show considerable morphological variability.

The Xianghualing mining area produced tens of thousands of specimens of calcite on fluorite, though many were not of the caliber of this 6.2 cm high piece ***(right)*** *in the collection of Paul Bernhard. Photo Jeff Scovil*

Below: *A wide variety of calcite habits colored green by malachite have come from the Dashunlong mine. This 5.8 cm wide cluster of flattened rhombohedra is in the collection of Terry Huizing. Photo Terry Huizing*

To the northeast, 45 kilometers south of the city of Hongyang and 30 kilometers east of the Changning County line, the **Shuikoushan lead-zinc mine** produced combinations of rhombohedral and scalenohedral calcite crystals, some with phantoms. The crystals were primarily light red, dark red, brown and pale pink. Recently, translucent and white, tabular, hexagonal crystals were also found at this mine. These had shallow pyramidal terminations, were sometimes associated with sphalerite and galena and were on a matrix of siliceous limestone.

In the Shuikoushan mining area, there are two important orebodies. A major mining target, the Shuikoushan deposit consists of metamorphic aureoles localized along the contact between granitic intrusions and Permian carbonates. The ore-bearing belts are tens of meters wide, are up to 1 kilometer long and are more than 1 kilometer deep. The primary mineralization includes sphalerite, galena, chalcopyrite, arsenopyrite, pyrite, hematite, scheelite, calcite and fluorite.

The other orebody, the Kangjiawan deposit, is relatively small, but its mineralization is similar to that found at Shuikoushan. Both deposits are thought to have formed as the result of contact metasomatism and hydrothermal activity. Calcite crystals are found as associated species at the Shuikoushan mine and are not as common as the sphalerite and galena that is found there.

Shangbao Pyrite mine, 30 kilometers east of Leiyang, is known not only for purple fluorite, pyrite and scepter quartz, but also for dolomite and calcite. Here, calcite crystals generally form as overgrowths on dolomitic or quartz matrix and are associated with fluorite, quartz and pyrite. Rosettes and doubly terminated hexagonal calcite are among the outstanding specimens that this mine has produced. The Shangbao is a pyrite-dominant, multiple-metallic deposit. Its ore-forming setting has been interpreted as the reaction between a carbonate host and intrusions of acid to neutral composition. Purple fluorite, large, lustrous pyrite crystals and scepter quartz are the most important specimens that the Shangbao mine has produced.

Southeast of the Shangbao mine, southeast of the city of Chenzhou, pink *manganoan* calcite crystals are found in two tungsten mines: **Yaogangxian** in Yizhang County and **Shizhuyuan** in Chenhzou. Crystals normally form in stacked, rounded and/or hexagonal thin plates. White, stacked, thin

Upper Left: *Calcite on dolomite, 6.5 cm high, from the Shangbao pyrite mine in Hunan. Collection Jeff Scovil*

Upper Right: *A 10.5 cm wide clear twin from one of the lead-zinc mines in the Babu area in Guangxi. Fine Minerals International collection*

Left: *These 7.5 cm high, transparent calcite crystals are from the Dachang mine in Nandan, Guangxi. Brown and Wilensky collection*

All photos Jeff Scovil

hexagonal crystals with lines of gold-colored inclusions of fine pyrite particles have been also found in these two mines. The Yaogangxian mine typically and frequently produces superb thin plates of calcite overgrowing quartz, wolframite, arsenopyrite, stannite and fluorite.

The Yaogangxian and Shizhuyuan mines are both well known as tungsten-dominated, multiple-metallic deposits. These orebodies, rich in mineral species and crystal habits, include a variety of genetic types ranging from granitic pegmatites, hydrothermal veins and contact metamorphic deposits. More than 50 mineral species have been reported from these mines. The most common minerals there are wolframite, scheelite, arsenopyrite, chalcopyrite, fluorite, bournonite, calcite, dolomite and stannite.

Seventy kilometers southwest of Chenzhou and 20 kilometers north of Linwu County, the **Xianghualing mining area** encompasses dozens of small tungsten and zinc mines and is quite famous for green fluorite in a number of habits and associated with an array of species. The mining area also produces a wide variety of calcite crystals: colorless and clear, tabular crystals and rosettes are typical. Calcite is typically associated with fluorite, enhancing the value of the specimens. Plates of calcite and fluorite can range from a few millimeters to more than 50 centimeters across. Over the last 8 years, the Xianghualing mining area has produced tens of thousands of specimens of calcite and fluorite. A series of tin, tungsten, lead-zinc and beryllium deposits have been developed in this area, in which calcite is an associated species. These deposits are thought to be products of hydrothermal intrusions into the Devonian carbonate and fine clastic host rocks.

There are many more notable calcite finds in Hunan, though of a somewhat lesser magnitude. These include light yellow, short pyramidal crystals from the **Xikuangshan Antimony mine** in the city of Lengshuijiang. This, the largest antimony mine in the world, is famous for superb

Karst Hills

This riverbank scene, backed by the dramatic karst landscape, was photographed by Rock Currier as he sailed down the Li River in 2000.

stibnite crystals. The same mine also rarely produced fantastic specimens of yellow, gemmy calcite intergrown with stibnite crystals. In western Hunan, yellow calcite and dolomite are widely found in cinnabar localities such as the **Chatian mine** in Fenghuang County. Recently, golden calcite crystals have been found in the **Lutang coal mine** near Chenzhou.

The Karst Hills of Guangxi

The Guangxi Zhuang Autonomous Region is another area rich in calcite. This is no surprise as more than half of the 230,000 square kilometers of this region have surface exposures of limestone. In fact, Guangxi is known for its natural features and karst geography. The region's most important tourist attractions are the karst hills and caves at Guilin and in the surrounding areas. Forty-six million minority peoples, predominantly Zhaung, populate this region which enjoys, at least in theory, some element of self-government.

Calcite speleothems are widely distributed in the northern part of the region, including Guilin, Yangshuo, Liuzhou and Hezhou areas. Though the Chinese government has imposed regulations prohibiting the sale of cave formations, farmers from these areas often collect stalactites and stalagmites from the caves to sell on the open markets or on the weekend stone markets in both Guilin and Liuzhou. For generations, cave formations have been carved and left in situ.

In 1998, amber-colored, translucent, double terminated calcite and twins on {0001} were discovered at **Xionghuang Kuang** (translated *realgar mine*), a private mine located in the Wuyu district of Hechi. These aesthetic and gemmy specimens were quite lustrous and were often found overgrowing short stibnite crystals. Many of the specimens in the initial lot were damaged during extraction, but after several weeks, the quality of the specimens improved as the miners learned how to properly collect them. Sadly, the last pocket was found late in 1999; still, the mine produced some 2,000 specimens of calcite on stibnite.

The miners' increased awareness of the value of calcite specimens has prompted new calcite finds around both Hechi and nearby Nandan. Red, yellow, black and colorless calcite crystals and groups have been found repeatedly over the last several years. The most important among the calcite finds are probably the large, clear, penetration twins, up to 25 centimeters across as well as the transparent, long hexagonal prisms. These were recovered at the **Dachang mine**, in the area of Nandan, just 10 kilometers north of the realgar mine in Wuyu.

Geologically, Hechi and Nandan share the same mineral-forming setting, the so-called *Dan-Chi multiple metallic mineral-forming zone*. This zone extends over 15 kilometers from Mangchang southeast via Dachang to Wuyu. It has provided various ores to a number of area mines. The deposits are dominated by antimony, silver, lead and zinc minerals in the northern section of the zone near Mangchang; by tin, silver, copper and arsenic minerals in the middle section near Dachang; and again by antimony, silver, lead and zinc minerals in the southern section near Wuyu. Antimony, tungsten, silver, mercury, bismuth, cadmium, gallium, indium and gold minerals and crystals of fluorite and calcite are also found in these deposits. The ore-bearing bodies are mainly associated with tensional faults occurring in Devonian marine carbonates and shales and with granitic intrusions during the Late Mesozoic. The mineral-forming processes are generally regarded as hydrothermal and as contact metamorphic activities that occurred in multiple phases and times.

Localities for calcite continue to spring up throughout the region as the market for mineral specimens develops. In 2002, beautiful, transparent, colorless calcite twins turned up from the **Babu area lead-zinc mines** near the city of Hezhou in Guangxi. Calcite is often found in association with the metals that are found there.

Other Provinces

There are other important calcite localities in China. For example at the **Daye Wollastonite mine** in Hubei province, clusters of scalenohedron scepters capped by flat rhombohedra, and partially colored red by hematite, were available at the 2001 Munich Show. Various habits of yellow, clear and colorless calcite were discovered last year. The **Qinglong mine** in southern Guizhou Province produced very nice transparent calcite crystals including penetration twins and needle-like clusters. The **Zunyi area** in Guizhou yielded white tabular calcite and brown rosettes.

China has also abundant resources of colorful massive calcite that is mined for carving and decoration. These deposits are distributed throughout a number of provinces, but most importantly in Hunan, Guangxi, Yunnan, Henan and Guizhou of southern China.

The privately held Xionghuang Kuang yielded this 5.9 cm high stibnite-included crystal that is now in the collection of Jeffrey Starr.

Photo Jeff Scovil

Specimen Mining and Trade

Mining and trade of mineral specimens really began in China in the late 1980s, just a few years after China opened its doors to western tourists. Before that time, all of the mines were government owned and operated, and there was no market for mineral specimens. Geological institutes and museums did collect a limited number of specimens for research and educational purposes.

As western visitors began to come to China, geological museums and institutions began to trade specimens. Collectors frequented the Hunan Provincial Museum and Geological Institute, in the capital city of Changsha, as the region was

Another stunning specimen from the 1999 find at Xionghuang Kuang. Steve Smale purchased this 5.8 cm calcite twin on {0001} on stibnite from a dealer in Hong Kong. He later sold it to Sandor Fuss, in whose collection it is now housed. The photo is by Jeff Scovil.

well known for fluorite, realgar, cinnabar and stibnite. These visits led to the establishment, in the early 1990s, of the first mineral shops in China. The museum and institute-owned stores focused on selling the duplicate specimens in their own exhibit or research collections.

In the middle of the 1990s, with the introduction of private enterprise in China, commercial mineral dealers began to replace the state-run shops. The first dealers had been previously employed in the state-run stores. Today, these dealers run the most exclusive shops in Changsha. In addition, some miners hung up their hard hats, moved to Changsha and became professional dealers.

In recent years, decorative stone and stone carving markets have sprung up in a number of cities; Guilin and Liuzhou in Guangxi may be the most important of these. Mineral specimens are also found in these markets. As in many places around the world, the US dollar is the currency of choice when buying mineral specimens in China.

It is very difficult to buy mineral specimens directly from the miners, as most mines are still government-owned, and miners are not allowed to sell minerals. In general, the miners have no knowledge of the mineral market and tend to quickly sell new finds to the professional Chinese dealers. Most of the mines are situated in undeveloped, mountainous areas and are accessed by poor roads. While a few large-scale mines employ modern mining techniques, most deposits are still mined simply with hand tools.

Despite the difficulties in accessing the mines or miners, southern China is worth visiting for the subtropical climate, unique vegetation, dramatic rivers and lakes and the spectacular karst caves. Many different minorities including the Zhuang, Miao, and Tujia populate Guangxi, Hunan and Guizhou, providing these areas with a colorful mix of culture and history.

RUSSIA
Lake Baikal
Irkutsk
Transsiberian Railroad
Khabarovsk
Ulaanbaatar
NEI MONGOL
HEILONGJIANG
ALTAI Mountains
MONGOLIA
Changchung
Dal'negorsk
Rudnaya Pristan
JILIN
Vladivostok
NEI MONGOL
Great Wall
Shenyang
LIAONING
XINJIANG
Bishkek
KYRGYZSTAN
NEI MONGOL
GOBI DESERT
GANSU
Huang He
Yellow River
Great Wall
Beijing
Tianjin
NORTH KOREA
Pyongyang
Dalian
TAKLIMAKAN DESERT
BEIJING
NINGXIA
SHANXI
Seoul
JAPAN
Jinan
Qingdao
SOUTH KOREA
QINGHAI
SHANDONG
GANSU
JAMMU & KASHMIR
Osaka
SHAANXI
JIANGSU
XIZANG TIBET
CHINA
HENAN
Nanjing
HIMACHAL PRADESH
ANHUI
Shanghai
HUBEI
PUNJAB
Chengdu
Wuhan
HIMALAYA
SICHUAN
Hangzhou
HARYANA
NEPAL
ARUNACHAL PRADESH
Chang Jiang
Chongqing
ZHEJIANG
Delhi
Yangtze River
JIANXI
Kathmandu
BHUTAN
HUNAN
UTTAR PRADESH
ASSAM
GUIZHOU
FUJIAN
NA.
GU.
Taipei
BIHAR
BANGLADESH
MA.
INDIA
YUNNAN
Xi Jiang
GUANGDONG
TR.
Dhaka
MI.
Pearl River
Guangzhou
TAIWAN
WB.
GUANXI
Hong Kong
MADHYA PRADESH
Calcutta
MYANMAR
Detail Map
Jalgaon
ORISSA
Hanoi
Nasik
LAOS
SI. SIKKIM
NA. NAGALAND
GU. GUWAHATI
MA. MANIPUR
TR. TRIPURA
MI. MIZORAM
WB. WEST BENGAL
MAHARASHTRA
Vientiane
Poona
Hyderabad
Yangon
ANDHRA PRADESH
THAILAND
GOA
Manila
Bangkok
VIETNAM
CAMBODIA
PHILIPPINES
KARNATAKA
Madras
Phnom Penh
TAMIL NADU
Ho Chi Minh City
KERALA
SRI LANKA
Colombo
Satellite Image Backdrop courtesy of NASA
Source: MODIS / Terra Satellite, 1km resolution
Projection & Datum: Geographic, WGS84
Compilation: MMC, Germany
S. Oeldenberger
MMC 2003
For Lapis International
This Map is not an Authority on National and International Borders

Topographic Map of the Hunan Province and the Guangxi Zhuang Autonomous Region, Southern China

Scale: *1:5,000,000*

Geographic Projection

Area Size: *approximately 1,000 x 900 km*

Legend:

⚒ indicates a mine mentioned in the preceding article (pages 48 to 54)

Cities underlined in red are reference points mentioned in the article

Backdrop © Robertson McCarta, Ltd (London, 1985)..

Calcite from the Deccan Traps of India

Dealer/collector ***Bertold Ottens*** *of Spiegelau, Bavaria, Germany has traveled more than 50 times in 20 years to India to study its geology and mineralogy. He summarizes here the information that he has collected on the calcite occurrences of the Deccan Traps.*

The Deccan flood basalts cover a 500,000 square kilometer area of west central India. These basalts stem from Cretaceous and Tertiary bubble-rich lava flows and reach 1,500 meters in thickness. They are predominantly tholeiite and are composed of some 10 weight percent CaO, a prime environment for calcite formation. Indeed, specimen-quality minerals are found at more than 120 quarries peppering the Maharashtra state; calcite is found in half of these. Calcite is thus widespread; still, it has only recently gained significance both among miners in India and among collectors, as both have focused primarily on India's colorful, lustrous zeolites.

The mineralization lining the cavities of the Deccan basalts was first discovered during the 19th century when railroads were built between Khandhala Ghat and Poona and between Kasara Ghat and Nasik. Many collectors became familiar with the Deccan minerals during the 1970s and 1980s when magnificent green fluorapophyllite in association with mesolite came from Poona, salmon-colored stilbite came from Nasik and aesthetic, well-developed okenite balls came from Mumbai*. The 1990s brought spectacular finds of blue cavansite from the Wagholi quarries near Poona, but it was the fluorapophyllite, stilbite and red heulandite from Jalgaon that gave rise to a central role for Indian minerals on the specimen market.

Quartz enhances, but fortunately does not hide this splendid, 5.9 cm high calcite crystal from the Mahodari quarry near Nasik.

Collection Bill Butkowski; Jeff Scovil photograph

Stepchild in the Mineral Family

All the while, calcite was also being found in the cavities of the basalts. This form-rich mineral, however, has only recently gained notice, as in most quarries calcite tended to occur as simple, whitish rhombohedra. Intense yellow calcite was rarely found, but when it did occur, it was often in association with heulandite, stilbite and fluorapophyllite.

The spilitic basalts of the **Malad/Kurar quarries** in the Mumbai area produced colorless to brownish, complex calcite crystals, aggregates of which present a spherical habit; nevertheless, calcite continued to be thought of only as an accessory to India's other mineral treasures. Finds in the early 1990s at the **Mumbai Dahisar quarry** finally brought calcite, in its own right, to the attention of the mineral market. While the quarry produced few zeolites, it yielded several generations of calcite in a wide variety of forms, including magnificent specimens of clustered scalenohedra and cherry-colored crystals in association with hematite that were highly prized as specimens.

Three Generations of Calcite

Throughout the Deccan basalts, calcite mineralized in three distinct generations. The first generation crystallized at relatively high (250 - 200 degrees C) temperatures on clay minerals and lining pockets. This generation is often found coated with the chalcedony, agate and amethyst varieties of quartz. A second generation formed after the quartz, but before the formation of India's famous zeolites. The third and final generation of calcite formed along with gyrolite, okenite, fluorapophyllite and aragonite after, and sometimes atop, the zeolites. This generation is fairly rare and is not common to the bulk of Deccan mineral specimens.

**In the mid 1990s the Indian government changed the name of the city from Bombay to Mumbai*

Color-Zoned

This striking 7.8 cm high zoned crystal came out of the Malad quarry in the Mumbai area and is now is the collection of Judy Megerle.

Photo Jeff Scovil

Most of the calcite specimens are found in the quarries around greater Mumbai area and in Jalgaon. The **Jalgaon-Sawda quarries** carry the first two generations of calcite. The first generation includes pseudoprismatic twins on $\{01\bar{1}2\}$, elongated along the distorted rhombohedra that reach 30 cm in length and that often exhibit multi-phase growth. Unfortunately, these crystals are typically coated with quartz. They are often also overgrown with heulandite, stilbite and fluorapophyllite, making it difficult to recognize the original crystal shape. The second generation crystallized as rhombohedra that are very often twinned on {0001}. The crystals are generally white, though they are occasionally intensely yellow, highly transparent and lustrous and partially overgrown with zeolites.

Other outcrops from time to time produce interesting and well-developed crystals and associations, such as specimens with intensely yellow rhombohedra from the **Pashan Hills** and **Wagholi quarries** in the Poona area. Mines in the Nasik-Shakur region including **Panduleni Hill**, **Eklara** and **Sinnar** quarries have at times produced third generation scalenohedra with steep faces and long, prismatic, needlelike crystals and encrustation pseudomorphs.

Calcite habits vary widely from simple, flat rhombohedra to steep, form-rich scalenohedra. Indian calcite can occur as tabular, long prismatic and pseudo-cubic rhombohedral crystals. Twins are most common on $\{01\bar{1}2\}$, but other twin laws are also found from time to time. Calcite is generally colorless and water-clear but yellow, brown, dark green and black crystals are found as well, some with pronounced color zones. In spite of the market's increased awareness of the significance of Deccan traprock calcite, specimens are less prized for their unusual forms and more for their aesthetics as accessory minerals.

Editor's Note: *Bertold Ottens has written a comprehensive article on the calcite occurrences of the Deccan Traps. This not-to-be-missed article is scheduled to appear in* Rocks and Minerals *magazine in 2004.*

Left: *The heulandite crystal is a feather in the cap of this lustrous, golden, gemmy crystal from Jalgaon. Fine Minerals International; overall specimen 11,4 cm high*

Right: *A pair of first-generation calcite twins from Jalgaon has been entirely encrusted by quartz. This 5.3 cm high specimen is in the collection of Terry Huizing.*

Both photos are by Jeff Scovil

Facing Page: *Certainly not an accessory here, this stunning 10.4 cm high calcite crystal from Lonavala is set off beautifully by a "bowtie" stilbite crystal. Photo and collection Steve Smale*

Dal'negorsk, Russia:

Terry Huizing with a whirlwind tour through the fabulous Dal'negorsk calcite localities

The mineral deposits at the town of Dal'negorsk, Primorskiy Kray in far eastern Russia have quite possibly produced the world's finest calcite. The clarity and transparency of the best Dal'negorsk calcite crystals are unsurpassed.

Here, a variety of beautiful crystals, with lengths to 70 centimeters (Moroshkin, 2001), occur in a number of habits involving long and short prisms, the pinacoid, and a variety of rhombohedra and scalenohedra; as twins on {0001} and $\{01\bar{1}2\}$; and with colors and clarity that range from colorless and transparent to translucent white, pink and brown-to-black. Calcite is frequently associated with other wonderfully crystallized minerals.

Above: The Port at Rudnaya Pristan' *Dal'negorsk ores made their way via rail through the port city of Rudnaya Pristan' and on to Japan and other world markets.*

Upper right: *A couple of helpers making a Primorskiy road momentarily passable*

Photos Ted Johnson

The mineral deposits at Dal'negorsk were exploited from the late 1800s, first by the Chinese, followed by the Russians, the British and once again the Russians. In the early 1900s, a railway was constructed to transport Dal'negorsk ore the short 35 kilometers to the city of Rudnaya Pristan' on the Sea of Japan and to world markets (Grant, 2001). Ultimately eight mines were developed to exploit the region's lead-zinc polymetallic ore deposits and the borosilicate deposits.

In spite of its proximity to the Sea of Japan, Dal'negorsk, with its population of 46,000, is both physically and culturally remote. Vladivostok, the region's largest city with 600,000 inhabitants and two international flights per week into its airport, is 330 kilometers southeast of Dal'negorsk. As infrastructure does not seem to be a government priority, however, it takes 8 to 10 scenic but uncomfortable hours to travel that 330 kilometers by car.

The Skarns Have It

Here in a karstlike environment, skarn bodies of hedenbergite composition host the sulfide and boron ores. Hydrothermal cavities, some to house-size, contain the minerals of interest to collectors. More than 160 mineral species have been reported from the Dal'negorsk area, and a baker's dozen, 13 species including calcite, are known to occur as good to outstanding crystals.

Remarkably, the region's mineral specimen treasures were largely unknown outside the Soviet Union until 1988, when Vladimir Pelepenko, one of Russia's leading collectors, displayed specimens from Dal'negorsk at the Munich Show. Then the rush was on! It is remarkable that specimens from the mines of the region have found their way into so many private and museum collections in but a few short years following collector recognition of the locality.

Calcite is one of the most abundant minerals in the Dal'negorsk deposits. Crystallization of calcite took place during several stages: during formation of the hedenbergite skarns at temperatures between 400° and 600°C; during the deposition of sphalerite and galena at temperatures between 120° and 350°C; and finally during the late chalcedony-quartz hydrothermal phase when temperatures ranged from 20° to 100°C (Grant, 2001). Late low-temperature calcite is often colorless and transparent; it frequently has a weak luminescence in ultraviolet radiation due to small amounts of Mn^{2+} substituting for calcium. By contrast, in some earlier generations the calcite is bright pink, translucent to transparent *manganoan*, and has very strong luminescence in ultraviolet radiation

Clearly Great Calcite

that is caused by the substitution of 1 to 2 percent Mn^{2+} (Moroshkin, 2001).

From the Dal'negorsk deposits in general, and the Verchniy mine in particular, the presence of up to four generations of calcite occurring in the same cavity is typical. The forms evolved in the following sequence: the $\{10\bar{1}1\}$ rhombohedron, the $\{10\bar{1}0\}$ prism, the $\{21\bar{3}1\}$ scalenohedron and finally the steep $\{02\bar{2}1\}$ rhombohedron. This generally follows the crystallogenetic diagram of calcite, according to Kostov (1968), and reflects the reduction of temperature during the crystallization sequence (Moroshkin, 2001).

Though calcite occurs in Dal'negorsk's borosilicate deposits including the **Danburity (Danburite) mine** and the **Bor (Boron) quarry**, it does not reach the same high standards as that from the polymetallic deposits. At the Bor quarry, in mineralized cavities to 10 meters length, combinations of the hexagonal prism and pinacoid are quite typical. Translucent yellow, to transparent colorless scalenohedra of calcite occur overgrowing quartz. Pseudomorphs of calcite after axinite (to 2 centimeters) and calcite after danburite (to 15 centimeters) have also been reported (Grant, 2001).

Left: *Ted Johnson captured this view of the Verchniy mine while on a collecting trip to Dal'negorsk in September 1997.*

Four Outstanding Producers

Of the six mines operated for polymetallic ores, two – the Sadoviy (Garden) mine and the Sentyabr'skiy (September) mine – have produced calcite of little collector interest; however, good to outstanding calcite has been produced from the other four mines.

Above: *The Verchniy mine is famous for water-clear calcite. This 7.6 cm wide specimen is in the collection of Francis Benjamin. Photo Jeff Scovil*

The **Verchniy (Upper) mine** was the first mine in the region, and it continues to produce excellent calcite specimens. The mine is known for its variety of calcite and for the large size of its crystals and crystal groups that occur mostly in cavelike openings in the rock.

The early generations of calcite exhibit a spear-shaped habit formed by a steep scalenohedron in combination with the prism. In practically all cases, this habit did not survive later growth episodes (Moroshkin, 2001). From later growth stages, remarkable doubly terminated, colorless and transparent to white scalenohedra, some terminated by a small rhombohedron, occur singly or as groups reaching 40 centimeters. Also notable are colorless, transparent, flattened crystals consisting of a short prism and scalenohedron terminated by a low, often striated rhombohedron or terminated by a combination of the rhombohedron and pinacoid. These crystals are sometimes associated with fluorapophyllite druse coating the calcite faces, and with ilvaite and quartz.

Twins on $\{0001\}$ and $\{01\bar{1}2\}$ occur as colorless, transparent scalenohedra to 8 centimeters across. A number of twins on $\{01\bar{1}2\}$, greatly elongated in a flattened prismatic habit and with the twin plane visible the length of the crystal, were collected in the late 1990s (Richards, 1999).

Large, translucent, obtuse scalenohedra of pink manganoan calcite that resemble the normal rhombohedron occur; these stubby crystals may be

Right: *In the 1990s, the Verchniy mine produced a large find of flattened calcite twins on* $\{01\bar{1}2\}$*; some were frosted and others were clear. The longest reached 15 cm. The specimen pictured is 7.1 cm high and is in the collection of Eric Asselborn; Jeff Scovil photo*

overgrown with smaller, colorless, short prismatic crystals with rhombohedral terminations.

The **First Sovietskiy (Soviet) mine** opened in 1934 and closed around 1965; thus, specimens are primarily available from older collections (Grant, 2001). Here the first calcite formed giant crystalline veins at the contacts between limestone and ore deposits. In cavities, later crystallization formed the finest calcite from this mine: the leaflike *paper spar*, white translucent crystals to 10 centimeters consisting primarily of the pinacoid. This habit was followed by prismatic crystals terminated by the pinacoid and by combinations of an obtuse rhombohedron and prism.

The **Second Sovietskiy (Soviet) mine**, also opened in 1934, is in partial operation at this time. Outstanding specimens of calcite are found in large limestone karst chambers to 50 meters in length. Calcite occurs as simple white crystals consisting entirely of the prism and pinacoid; white tabular crystals with a prominent pinacoid termination; simple tan rhombohedra associated with galena; and colorless, transparent prisms, both thick-tabular and columnar, each terminated by a low rhombohedron. Some prisms are associated with ilvaite and hedenbergite, others occur twinned on $\{01\bar{1}2\}$.

Translucent and often bright pink manganoan calcite occurs as flat rhombohedra overgrowing normal scalenohedra that produce pagodalike crystals. Associations include a different generation of white, non-manganoan calcite as simple prisms and pinacoid as well as green, hedenbergite-included, cubic fluorite to 4 centimeters on edge.

The **Nikolaevskiy (Nikolas) mine**, opened in 1982, is now the largest producer of lead and zinc in the region. There is quite a contrast in calcite habits from this mine. Most specimens occur as crystal aggregates consisting of an obtuse rhombohedron that forms near-spherical intergrowths or saddle-shaped tabular crystals, many with siderite and pyrite overgrowths (Moroshkin, 2001).

Well-formed calcite crystals are widespread within the deposit. Their habit is prismatic and ranges from thick-tabular to columnar; they may be terminated by either an obtuse rhombohedron or by a combination of this rhombohedron and obtuse scalenohedron, which sometimes forms a pseudopinacoidal surface (Moroshkin, 2001). Crystals are typically colorless and transparent. At least one outstanding occurrence produced a transparent columnar crystal 11.8 centimeters long (below) in which hollow tubelike inclusions of aragonite can be seen throughout the crystal.

The pink manganoan variety of calcite typically occurs as stout prismatic crystals with rhombohedral terminations. As is typical of most mines in the Dal'negorsk region, some calcite luminesces orange-red under ultraviolet radiation from Mn^{2+} substituting for calcium.

Right: *The Nikolaevskiy mine has produced a number of interesting calcite habits, including the specimen pictured here: an 11.8 cm long crystal with tubular aragonite inclusions. Collection Terry Huizing; Jeff Scovil photo*

The Second Sovietskiy Mine

4

3

5

2

1

6

The Second Sovietskiy Mine and Its Calcite

1. *Photo of the mine entrance taken by Ted Johnson in 1997.*

2. *This 10 cm high calcite specimen found in 2000 at the 135 meter level of the mine is in the author's collection. Photo Jeff Scovil*

3. Mike Sheasley squeezes into a large vug. Photo Ted Johnson

4. *This 9 cm wide calcite specimen is included with fibrous aragonite. Collection Judy Megerle; photo Jeff Scovil*

5. *The author photographed and owns this 7x9 cm calcite and galena specimen that was mined in March of 1999.*

6. *"Poker chip" calcite is just one of the many highly-sought habits found at the Second Sovietskiy. The 7.3 cm high specimen is in the collection of Judy Megerle. It was photographed by Jeff Scovil.*

7. *Boxer turned miner, Dal'negorsk native Alexander Zaytsiv (right) poses with a friend and one of the day's finds. Photo Ted Johnson*

7

Cumbria:

England Is Known For...
Calcite and Calcite Twins

Upper Left:
Cumbria, England:
6.4 cm high, hematite-coated scalenohedral crystals
Martin Zinn collection
Jeff Scovil photo

•

Upper Right:
Pallaflat, Cumbria, England
14.5 cm high cluster of twin crystals on $\{01\bar{1}2\}$
Eric Asselborn collection
Jeff Scovil photo

•

Lower Left:
6.5 cm wide group of calcite crystals from the Bigrigg quarry in Egermont, Cumbria; Tim Sherburn collection; Jeff Scovil photo

The Legendary Calcite Twins

***Michael P. Cooper** from the Nottingham Museum in England on English calcite and its classic twins*

This "butterfly" twin from Cumbria has fine inclusions of hematite that give the specimen its color. It is 5.7 x 4.5 cm and is in the collection of Lindsay Greenbank who also provided this photo.

"Why should we vex our vision with microscopical crystals from Utah and Siberia, when there is so much glorious carbonate of lime to be had? The mineral dealer says, 'Here is a new species, you must keep up to the schedule,' and the customer blinks his eye and places another chunk of 'faith' in his cabinet. A rhomb of Iceland Spar is a joy forever. A crystal of native copper inside a crystal of calcite is an inclusion worth looking at. Bring on some more good calcites!"

When in 1896 American mineral collector Charles Pennypacker (1845–1911), eccentric columnist for the American journal *The Mineral Collector*, wrote those words, the mineral market was replete with superb calcites from English mines. Those from the lead mines of the Northern Pennine Orefield were typically white, short to long prisms with low rhombohedral terminations. Calcite twins were well represented by the cream to golden brown crystals taken from vugs in Derbyshire. Even Cornwall, where crystallized calcite is surprisingly rare, yielded a plethora of habits ranging from thin, pale plates of delicate *paper* or *angel wing* calcite (typical for West Cornwall) to the classic but rare twins on $\{10\bar{1}1\}$ from Wheal Wrey. It was the calcite from the hematite deposits of West Cumbria (then known as *West Cumberland*), however, that dominated the imagination of mineral collectors and swelled the inventory of many a mineral dealer the world over.

West Cumbria

The West Cumbrian iron mines, and those of the Furness District of what was then northern Lancashire, worked some of the richest hematite deposits in the world. Their ore was highly prized in steel making by the then new Bessemer process. The deposits occurred in bedded limestone and contained many cavities, known locally as *loughs*, within which occurred a quantity and quality of calcite and barite never before known. In some cases these minerals were so loosely

Setting the Standard for Beauty

attached to the matrix that the miners could pluck the specimens from crystal-lined nests with their bare hands. The calcite crystals were unsurpassed and quickly set a standard for beauty that has been rarely matched since.

The best of these limpid, glassy, multi-faceted jewels reached lengths of 10 centimeters. They occurred in sculptural radiating groups or as loose single crystals exhibiting an amazing number of forms that combined various scalenohedra, prisms and rhombohedra with a profligacy that is unmatched in other occurrences. As if this were not enough, crystals were also found in a stunning variety of twinned habits. Crystals from some occurrences are enhanced by a glaze or by inclusions of rust-red hematite (typical of specimens from Furness, but also known from Frizington and elsewhere) or by delicate traceries of black manganese minerals. The lower levels of workings at Bigrigg mine (especially the Whyndam pit) were particularly rich in manganese, even producing some crystals of hausmannite, a most unusual occurrence in England.

The Riber mine in Matlock, Derbyshire, England produced this 6 cm high scalenohedron twinned on {0001} with zoned pyrite inclusions. Bob King collected this specimen in 1957, and it is now part of the National Museum of Wales in Cardiff (King Collection) K1391 (NMW M5087). Michael P. Cooper photo

Henry Miers and the Twins

Specimens began to appear in 1888, and Henry Miers (1858-1942) from the Department of Mineralogy at the British Museum (Natural History) was quick to recognize their importance, describing them to the Mineralogical Society in 1888. The paper was published the following year. Research into the dates of workings suggests that these early twins came from mines at Pallaflat and Gilfoot. Miers continued to purchase specimens for the Oxford University Museum after he became Professor of Mineralogy there in 1895. Miers noted that all four calcite twin laws (see page 10) were represented by these specimens.

1. Twins on the pinacoid {0001}

The easiest twin to recognize is that on the pinacoid {0001} where both crystals share a common main axis and there is an equatorial plane of symmetry indicated by reentrant angles around the middle of the crystal. From their common occurrence among the crystals of Derbyshire, this scalenohedral habit is sometimes referred to as the *Derbyshire twin*. In crystals from Cumbria, the basic form is commonly the prism.

2. Twins on the positive rhombohedron $\{10\bar{1}1\}$

The top prize for Cumbrian calcite collectors is the twin on $\{10\bar{1}1\}$, in which the *c*-axes are inclined at an angle of 90°46'. In Cumbria, this twin law represents the majority of twin specimens, unlike other localities worldwide where it is most uncommon. These twins are identified by the existence of a cleavage plane in each crystal that lies parallel to the composition plane. The overall habit of these specimens depends on the basic form of the twinned crystals and the extent

continued on page 70

Four Twin Laws, Numerous Habits

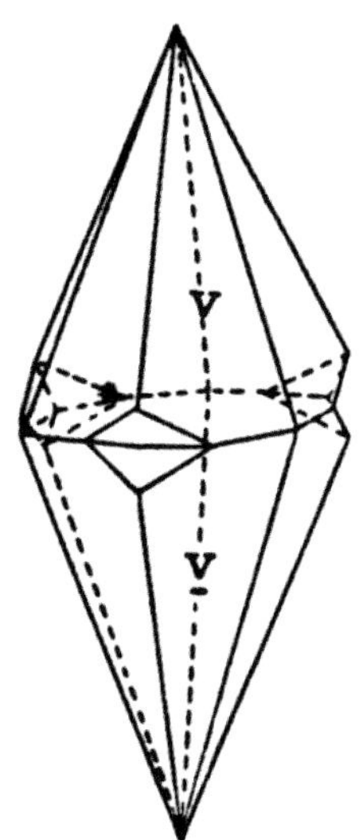

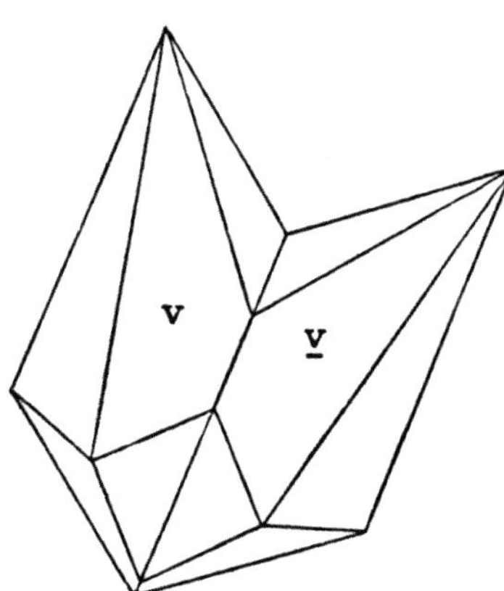

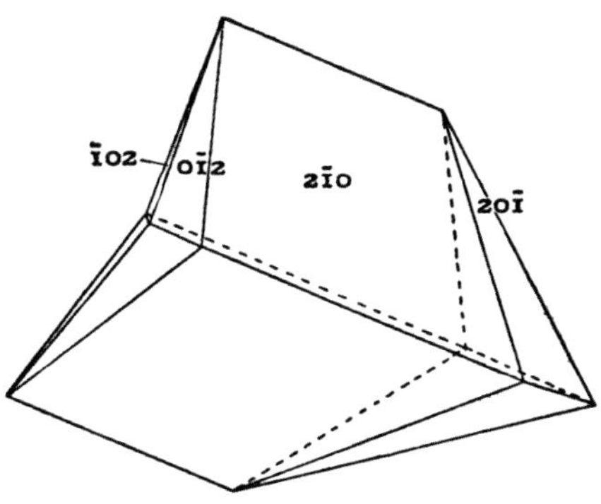

Cumbrian calcite twins as drawn by Henry Miers, London 1902

Left: *A scalenohedron that is twinned on the basal pinacoid* $\{0001\}$. ***Middle***: *A scalenohedron that is twinned on the steep negative rhombohedron* $\{02\bar{2}1\}$. ***Right:*** *A crystal twinned on the shallow negative rhombohedron* $\{01\bar{1}2\}$; *the indices used in this drawing follow a convention that is no longer used.*

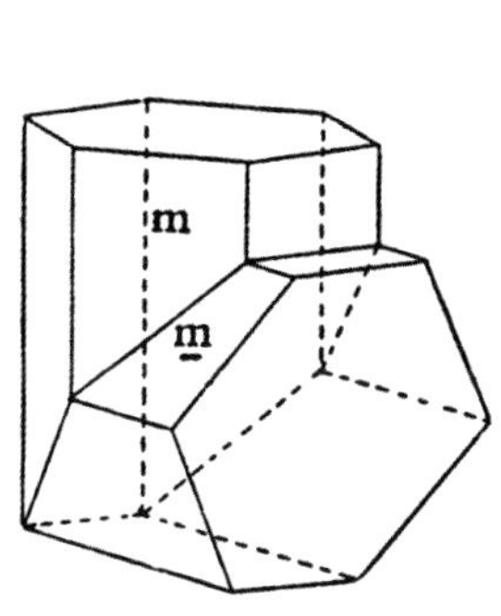

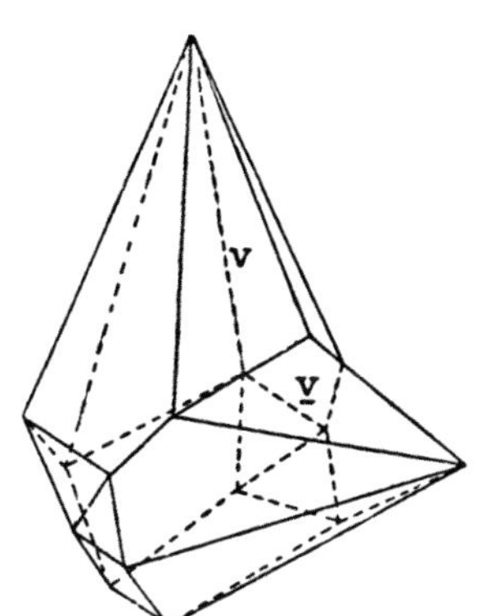

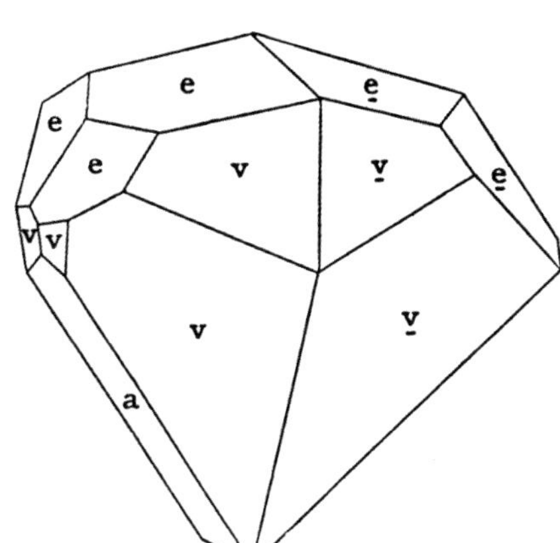

Above: *Three very different looking twins all of which follow the same twin law: they are all twinned on the positive rhombohedron* $\{10\bar{1}1\}$. *The apparent differences are a consequence of the differing habits of the crystals.*

Drawings by Henry Miers, London, 1902

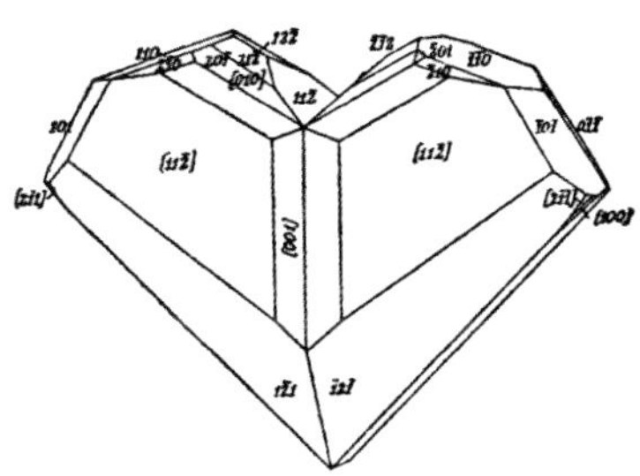

Many mineralogists have taken an interest in twinned calcite from Cumbria. Above and to the right are two drawings of calcite twinned on the positive rhombohedron $\{10\bar{1}1\}$ *from Egremont taken from Victor Goldschmidt's* Atlas der Kristallformen *(Atlas of Crystal Forms). These were first published in a work by Kreutz in 1907.*

A Feast for the Eyes,

but an ambitious specialty for the collector: twins from Cumbria. Unfortunately these finds from the English iron mines belong to the past. Old pieces, however, do sometimes come with wonderful old labels and provenances.

A Rare Twin from Frizington in Cumbria, England
An 8 x 5.5 cm pair of transparent, colorless, prismatic crystals twinned on {10$\bar{1}$1} with low rhombohedral terminations. This specimen was once in the private collection of Richard Barstow (1947-1982); its previous history is unknown, but it is undoubtedly an old piece. It may have come, with others, from the collection of David Corse Glen in exchanges Barstow is known to have made with the Kelvingrove Museum. Ralph Sutcliff bought the piece in 1982 from Richard's widow Yvonne. It was subsequently acquired along with the remainder of the Sutcliffe collection by Lindsay Greenbank in 1991 (Cat. No. LG 29).

Frizington, Cumbria, England
A 5.8 x 4.5 cm twin crystal on {10$\bar{1}$1}; John Graves sold this piece to the British Museum of Natural History in 1901 for £1 (BM85593). It was part of a 1985 exchange with Ralph Sutcliffe for a Wheal Jane ludlamite, and was acquired by Lindsay Greenbank in 1991 (Cat. No. LG34).
Both photos this page Michael P. Cooper

Above: Egremont, Cumbria

A remarkable group of five colorless "butterfly twins" on $\{10\bar{1}2\}$*; the largest is 6 cm across. The 11 x 8 cm group is perched on a thin matrix. These twins are generally seen alone or off-matrix; groups such as this are most uncommon. An old German label with the specimen describes it as "selten schöne xx Stufe!" - beautiful and rare. This piece once belonged to German collector and sauerkraut manufacturer K. Rau who specialized in high quality aesthetic specimens. On his death in the 1970s, the results of his 50 years of collecting were split: his Saxon minerals went to the Dresden collection, and the remaining world-wide material was acquired by the Humboldt University in Berlin. This specimen was no. 796 in the University collection and was 922/87 in the Rau Collection. The Mineral Gallery of Frankfurt acquired it from the University, then sold it to Simon Harrison, who in turn sold it to Ralph Sutcliffe in 1986 from whence it was sold to Lindsay Greenbank (Cat. No. LG45). Photo Lindsay Greenbank*

Bigrigg Classic

Left: *Dealer / collector Herb Obodda has had this 8.5 cm high twin on* $\{10\bar{1}1\}$ *in his collection for years. It has an old John Sinkankas number on the back (JS/CA34). Numerous similar twin crystals came from the Croft pit in the Bigrigg quarry, Egremont, Cumbria, England.*

Photo Jeff Scovil

English Calcite:

Minerals of England,

AND WHERE TO OBTAIN THEM.

THE ENGLISH TWINS.

Very fine and rare specimens of every size and form. Twins 1 lb. in weight and over, numbers to select from. My *Private Collection* of these is the *finest and best in the world.* I secured all the Egremont Twins and have still the finest of them in stock. This splendid collection is now for disposal; a grand opportunity for Museums and those w shing to obtain the best specimens in existence of this class. Phantom crystals, Angle crystals, Triplets, etc.

JOHN GRAVES, Mineral Collector,

FRIZINGTON, CUMBERLAND, ENGLAND.

A section of an advertisement, circa 1895, for the sale of John Graves' mineral collection; he built a considerable collection very quickly, then offered it for sale. Unable to interest the European museums in the collection, he exported it to the USA.

to which the reentrant angle is filled. The simplest habits are twinned scalenohedra or prisms (the latter being rarer) with no infill. When the reentrant angle is filled we see two basic habits, depending on the relative development of the scalenohedron and rhombohedron. The former develops a sharp-edged habit that is sometimes known as a *butterfly* or *axe head* twin; the rhombohedral form develops the chunkier *heart twin*, which is generally found alone with barely a mark to show where it had been attached to the matrix. *Butterfly* twins are sometimes found on matrix.

3. Twins on the negative rhombohedron $\{02\bar{2}1\}$

Among Cumbrian crystals, twinning most rarely occurs on $\{02\bar{2}1\}$ in which the *c*-axes of the two individuals diverge at an angle of 53°46' to one another. This twin law is distinguished by the lack of parallel cleavage planes of the two individuals.

4. Twins on the negative rhombohedron $\{01\bar{1}2\}$

At other localities, a common twin law is on $\{01\bar{1}2\}$, where the *c*-axes are inclined at 127°30' to one another. This twin law is not well represented in Cumbrian specimens.

John Graves - Calcite Twin Specialist

For many of these wonderful specimens, we are indebted to the devotion of one man: John Graves (1842-1928), an iron miner from Frizington, in the heart of Cumberlands's iron-mining districts. His interest in minerals began sometime in the 1870s. When he took up professional mineral dealing in the early 1880s, he quickly became the foremost mineral dealer in West Cumberland and many of the finest specimens from the area passed through his hands. He was the principal supplier of West Cumbria specimens to the American dealers George Letchworth English (1864-1944) of Rochester, New York, and Alfred E. Foote of Philadelphia (the home of mineral collecting in the USA).

Graves' personal collection of calcite twins from Cumberland was the largest ever assembled. Despite its undoubted quality and his many attempts to interest buyers at home, however, a large proportion of the collection appears to have gone to the United States, bought around 1821 by George English for Ward's Natural Science Establishment of Rochester, New York. By the time of Graves' death in 1928, the mines were declining, and the mineral specimen production was waning. Since that time very few finds have been made, but West Cumbrian calcites remain among the most desirable and scarce mineral specimens, one of the great classics of mineral collecting.

Fortunately, West Cumbrian twins are not the only calcite crystals that England has produced. The iron-rich mines of Cumbria yielded superb untwinned crystals. In addition, exceptional specimens are still found in provinces such as Wales and Derbyshire and of course Cornwall.

Cumberland Twins and More

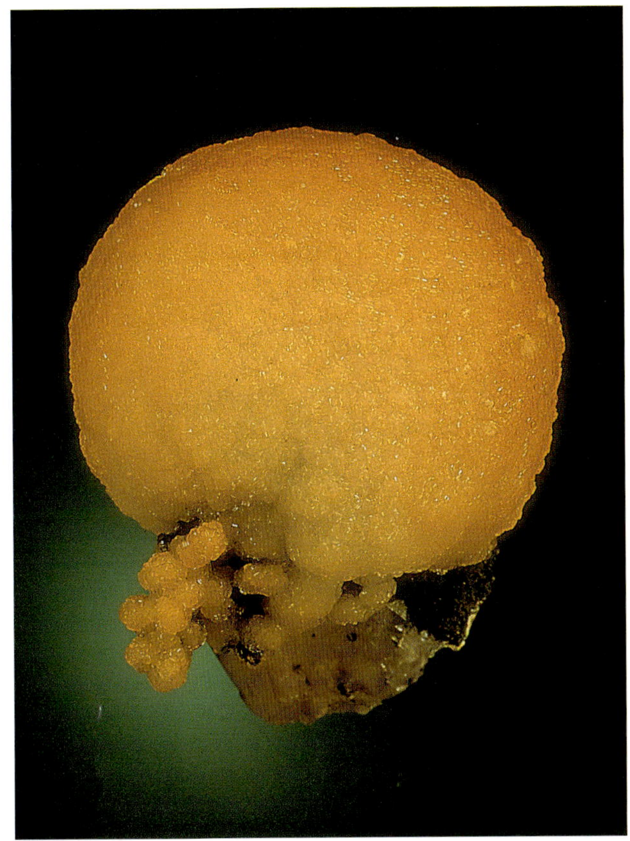

Upper Left:
A 4.7 cm high calcite from Wheal Cock, Botallack in Cornwall; collection Eric Asselborn; Jeff Scovil photo

Upper Right:
A rarity from Wales: a 6 cm tall crystal from Taff's Well in Cardiff, Mid Glamorgan, Wales; collection Ian Jones; Michael P. Cooper photo

Lower Left:
A 7 cm wide twin on {0001} from the Ladywash mine in Eyam, Derbyshire; Derby Museum collection; photo Michael Cooper

Lower Right:
A 3.8 cm high calcite and fluorite from Cumbria. Martin Zinn collection Jeff Scovil photo

Saint Andreasberg in the Harz Mountains:

Postcard collector, economic geologist and assistant curator of the Bavarian State collection (Munich) ***Günter Grundmann*** *shares his fondness for the legendary Saint Andreasberg silver deposits in the German Harz Mountains.*

The Harz is one of the oldest mining regions in Europe, dating back to the Bronze Age (circa 1500 B.C.). This mountain ridge, in the center of Germany, is home to historic districts such as Bad Grund, Clausthal-Zellerfeld, Neudorf, Rammelsberg and Saint (in German, *Sankt*) Andreasberg. The silver mines of Saint Andreasberg rank among Germany's greatest mineral localities. For the last three centuries, the world's museums and collectors have considered specimens from Saint Andreasberg to be among their greatest treasures. The words *Saint Andreasberg* have become synonymous with first-class calcite crystals as well as fine silver ores including pyrargyrite and dyscrasite.

Circa 1900

Two "chromo-lithographie" postcards from the author's archive: the upper of the two shows the shaft building of the Catharina Neufang mine to the left of the miner. The village of Saint Andreasberg is pictured to his right. The lower postcard depicts the city's coat of arms.

Right: *Judy Megerle owns this 2.2 cm high specimen of Saint Andreasberg "canon spar." Photo Jeff Scovil*

Bonus from the Silver Mines

Early records indicate that silver mining near Saint Andreasberg began in 1487, but the period between 1700 and 1900 was the locality's golden age for mineral specimens. Even René Just Haüy, the famous French mineralogist, owned and described numerous calcite and silver ore specimens from Saint Andreasberg. Well-known mines, shafts or veins are *Andreaskreuz*, *Juliane*, *Felicitas*, *Samson*, *Franz August*, *Jacobsglück*, *Gnade Gottes*, *Fünf Bücher Mosis*, *Catharina Neufang* and *Bergmannstrost* . On 30 March 1910, as a casualty of a dramatic decline in the silver market, the world-famous Samson mine became the last in the region to close its doors. Recent small-scale mining activities by collectors have consisted of mucking out the old workings and drilling by hand for specimens.

The hydrothermal lead-zinc-copper-silver vein deposits in the Harz Mountains are linked to northwest to southeast trending shear zones. The country rocks are generally sandstones, limestones, shales and volcanics of varied chemistry. The metals lead, zinc, copper, silver and subordinately antimony, arsenic, cobalt and nickel were leached from deep levels of the Paleozoic sediments and enriched in faults and shear zones. Mineralization is estimated to have occurred at temperatures between 250 and 300 degrees C.

Calcite is by far the most abundant gangue mineral. Some veins are filled purely with calcite and are locally known as *reine Spatgänge* (pure calcite veins). Multistage fracturing and the subsequent circulation of metal-bearing fluids provided conditions favorable for the formation of gangue and ore minerals. Five major depositional events produced a characteristic sequence of mineral paragenesis with complex crystal-zoning, alteration and replacement.

Habit Forming

In a 1978 article, Professor Albrecht Wilke wrote that at least 144 individual calcite crystal forms, 391 combinations and all four calcite twin laws were described by the famous crystallographers Sansoni, Goldschmidt and Luedecke as having been found in the mines at Saint Andreasberg (Wilke, 1978). In a 1989 personal communication, Wilke later reported that the locality has produced more than 1,500 distinct habits of calcite.

More Than 1,500 Distinct Habits

The largest calcite cavity ever found in the region was 10 meters across and 1 meter high. It was discovered in the *Fünf Bücher Mosis* vein and produced calcite crystals to 25 centimeters across. The largest calcite crystal found in Saint Andreasberg measured 1 meter across.

Miners coined a number of terms to describe some of the great variety of Saint Andreasberg calcite crystals (calcite has historically been referred to as *Spat* in German and *spar* in English). Literally translated, these terms include *needle spar*, *leaf spar*, *honey spar*, *pig's teeth*, *paper druse*, *atlas spar*, *spindle spar*, *lens spar*, *cubic spar*, *pencil spar*, *canon spar* and *pin spar*.

Calcite crystals range in color from transparent and colorless to milky white, green, yellow, orange, pink and brown to black and opaque. A characteristic feature of calcite from Saint Andreasberg is the occurrence of brown patches, especially at the cores of the crystals. Multistage growth is sometimes documented in the calcite crystals by zonal structures and phantoms. Classification of calcite generations in Saint Andreasberg calcite is enabled by the colors in which they fluorescence. Early generations of calcite fluoresce a deep shade of red, while younger generations fluoresce in hues of green to yellow. Some calcite phosphoresces intensely, its afterglow lasting up to 2 seconds.

Above: *The deposits at Saint Andreasberg were mined primarily for silver ore, but the many forms of calcite were a boon for specimen collectors. This 5.8 cm high specimen is in the collection of Terry Huizing, who is also credited with the photograph.*

Elbingerode

Of the 2,544 habits of calcite documented in Goldschmidt's *Atlas der Krystallformen*, more than 500 are found at the Elbingerode deposit, situated 20 kilometers east-northeast of Saint Andreasberg in the Harz Mountains. Both of the deposits mined in the villages surrounding Elbingrode are distinct from those at Saint Andreasberg. Calcite is common in both deposits, but it is the pyrite mine, which produced the raw material for sulfuric acid, that is famous for magnificent calcite specimens. Sadly, finds from this mine are not well represented in either private or museum collections.

Below: *The pyrite mine at Elbingerode yielded this rare 7.8 cm high calcite twin on calcite. Collection and photo Steve Smale*

Calcite from Romania:

Marc L. Wilson, head of the mineral section at the Carnegie Museum of Natural History in Pittsburgh, Pennsylvania, traveled recently to eastern Europe, gaining first-hand knowledge of Romanian localities and acquiring specimens for the museum's suite of Romanian minerals.

The mines of Romania have recently produced a large number of high-quality calcite specimens. Most have been carefully collected, and beautiful specimens are available to suit every taste.

The Maramures District

The polymetallic vein system of the Maramures mining district in northern Romania has been known as a source of beautiful and interesting specimens of calcite for more than one hundred years. Associated mineral species are numerous, most commonly consisting of stibnite, quartz, chalcopyrite, pyrite, siderite and rhodochrosite.

Flattened rhombohedral crystals are the most common form of calcite currently available from the district. Calcite from the **Herja mine** in Baia Mare typically occurs as curved crystals to 4 centimeters across and as spherical, saddle-shaped, scalenohedral or stalactitic aggregates of smaller crystals. Color ranges from white, when pure, to dark gray or black when included with minute, acicular crystals variously described as boulangerite or jamesonite. Spectacular specimens result when white and black areas are included within the same crystals or aggregates. Of special note are spherical specimens divided into white and black hemispheres (facing page).

A particularly interesting specimen in the museum's collection consists of a complete sphere of white calcite, 2.4 centimeters in diameter, sparsely sprinkled with minute acicular crystals of an as yet unidentified metallic species encased in microscopic tan calcite crystals. The sphere is hollow and is filled with water. An air bubble is visible through the surface (enhydro). A similar specimen on a matrix of small quartz crystals reveals two concentric spherical shells of calcite, each approximately 2 millimeters in thickness.

Manganoan Calcite

Most of the calcite specimens from the Maramures district contain at least some manganese. These often exhibit pale fluorescence in delicate red or orange. The manganoan variety of calcite is most prevalent in the Cavnic area. Many of these specimens are highly fluorescent, though the effect is diminished in those that are coated with a later generation of nonfluorescent calcite. Flattened rhombohedral crystals similar to those from the Herja mine have been produced from the **Boldut mine** in Cavnic. These, however, are generally smaller and lack the boulangerite or jamesonite inclusions common at Herja. Attractive specimens of manganoan, white, flattened crystals in scalenohedral aggregates reaching more than 2 centimeters in length, partially encrusting sprays of splendent stibnite crystals to nearly 14 centimeters in length, were produced in 2002.

Large scalenohedral crystals of a delicate pink color; smaller, white crystals associated with lustrous pyrite; and creamy white spheres perched on a matrix of white quartz crystals are among the more notable calcite specimens recently produced from the Boldut mine. Drusy coatings of white to pale pink crystals are also noteworthy as are rosettes associated with rhodochrosite and quartz.

Author Marc Wilson (center), mineral dealer Ross Lillie (right of center) and geologist Andrej Gorduza (behind Ross) mingle with the miners underground at the Herja mine. Photo Jeff Scovil

A saddle-shaped aggregate of calcite found at the Herja mine. It is 8.8 x 6.5 x 5.2 cm and is in the collection of the Carnegie Museum of Natural History (Cat. No. CM28140). Photo Debra Wilson

Breathtaking in Black and White

Manganoan calcite occurs at the **Sasar mine** near Baia Mare as plates of large, pale pink, equant crystals. The habit of these crystals consists of a scalenohedron truncated by a rhombohedron.

Other localities within the district have produced fluorescent manganoan calcite crystals. The **Turt mine** near Satu Mare has recently produced lovely rosettes and spheres of white and semitransparent, flattened rhombohedral crystals on pyrite, sphalerite and siderite. Excellent specimens of white calcite on stibnite occurred at the **Breiner mine** in Baiut during the 1970s.

Among the specimens in Terry Huizing's collection is this 9 cm wide calcite rosette on siderite from the Turt mine.

Jeff Scovil photo

Pseudomorphous Calcite

Fine pseudomorphs of calcite after other mineral species have also been recently found in the Maramures district. From the Herja mine, calcite has replaced acicular crystals of unknown identity. The specimens consist of slender, curved, hollow casts of minute, tan or gray calcite crystals with minor stibnite. The casts reach lengths of 11 centimeters and occur as delicate intertwined groups.

Also from the Herja mine, the Carnegie has a remarkable specimen of calcite after the selenite variety of gypsum. The specimen, mined in the 1960s, consists of creamy buff-colored calcite as a hollow cast of a doubly terminated gypsum crystal 12.9 centimeters high. The crystal stands upright within a thin shell of white calcite and minute, acicular crystals of boulangerite or jamesonite. Other smaller pseudomorphous crystals accompany the main crystal.

Mined in the 1960s, this 12.9 cm encrustation pseudomorph of calcite after gypsum (var. selenite) from the Herja mine is part of the Carnegie Museum's collection (CM 27895).

Photo Debra Wilson

From the Boldut mine come replacements of tabular calcite crystals by finely crystallized, pale pink manganoan calcite. In some specimens, continued growth of manganoan calcite along the *c*-axis of the original crystals has resulted in interesting, waferlike aggregates. These crystals occur on a matrix of quartz crystals and are associated with small crystals of pyrite and chalcopyrite.

Other Localities

Calcite specimens from other Romanian mining districts are far less plentiful than from the Marmures district. Of note are large scalenohedral crystals of manganoan calcite on a matrix of slender quartz crystals from the **Ocna de Fier mine** in Banat. Specular hematite is occasionally an associated species. Originating in a skarn deposit, the calcite crystals are light buff in color and are highly fluorescent bright orange-red. The Carnegie specimen measures 13 centimeters in height. Specimens from the nearby **Ruschita mine**, also in a skarn deposit, consist of scalenohedral crystals, often in parallel growth groupings, associated with quartz and galena.

The Herja mine produced a number of black and white spherical aggregates such as this 4.3 cm wide specimen, which is in the collection of Francis Benjamin.

Photo Jeff Scovil

A Bountiful Harvest of Calcite

Lifelong calcite collector ***Terry Huizing*** *reports on the most important calcite deposits in the United States*

The finest calcites in North America are found in the midcontinent region of the United States, where large colorful crystals occur in Mississippi Valley–type deposits in aesthetic association with other well-crystallized minerals such as galena, sphalerite, fluorite and barite; where beautiful crystals are found in limestone quarries located along the flanks of structural basins; and where limpid crystals, sometimes with included copper occur in the Michigan copper deposits (page 83).

No other region on the continent can boast the same diversity of habit, stunning associations and variety of twins and inclusions.

An 18.3 cm high calcite "scepter" from the Shullsburg mine in Shullsburg, Wisconsin. Lance Hampel collection; Jeff Scovil photo

Mississippi Valley–type (MVT) Deposits

What are MVT deposits? They are mineral deposits with common characteristics that were first recognized as such in the United States. Midwestern MVT deposits are hosted in shallow-water carbonate rocks of Paleozoic age, mostly in dolomitized areas; each mining district is located on a geological dome or arch where mineralization is typically confined to certain stratigraphic rock units. The mineralogy is simple, consisting of sphalerite, galena, fluorite and barite with quartz and carbonates, such as calcite, plus traces of asphaltic material. Deposition took place in open spaces at low (100° - 200° C) temperatures from highly saline brines. Geologists have now classified other deposits and mining districts in other parts of the world as belonging to this type.

The author wishes to thank Daniel Stewart for his careful reading of this text

MVT deposits in the midwestern United States have long been the source of many fine calcite specimens. Other minerals that are common to these deposits are galena, sphalerite, fluorite and barite along with minor amounts of chalcopyrite, pyrite, marcasite and quartz. When sphalerite and galena oxidize, they form sulfates with no concomitant production of sulfuric acid; thus the common association of these two minerals produces a preservative environment for calcite (Eriksson, 1995). When iron sulfides predominate, however, post-mining oxidation may damage calcite. In the midcontinent region of the United States, there are five MVT localities that have produced noteworthy calcite specimens.

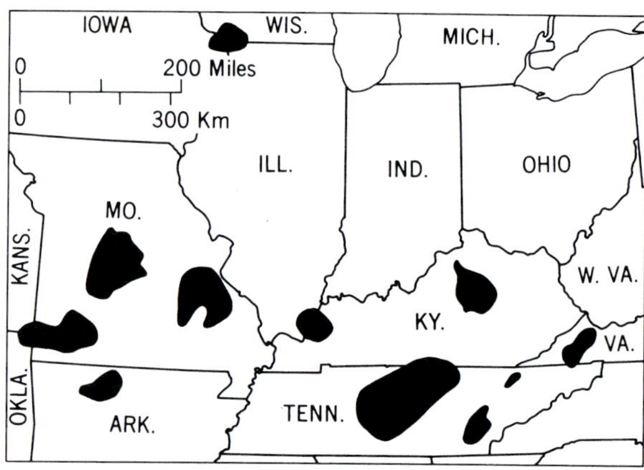

The midwestern United States with the Mississippi Valley–type (MVT) mineral deposits in gray Indiana Geological Survey Special Report 21

Upper Mississippi Valley Zinc-Lead District

This formerly important district is located in the southwestern corner of Wisconsin and extends to Galena, Illinois and Dubuque, Iowa. Operations for lead were recorded from as early as 1658, and a map from 1687 shows lead mines near the town of Galena. This district has the longest recorded production history of any mining area in the United States (Lasmanis, 1989). Following a decline in lead production, a zinc mining boom began that peaked in 1916 as a result of World War I demand. During the life of the district, more than 280 mines were developed in orebodies located in limestones of Middle Ordovician age (Lasmanis, 1989); most were in shallow deposits. By 1979 the last remaining mine at Shullsburg, Wisconsin had closed and was allowed to flood.

Many beautiful calcite crystal groups were

The Illinois – Kentucky fluorspar districts (Lasmanis, 1989)

recovered from these mines, though matrix specimens are difficult to preserve because the associated marcasite breaks down over time, releasing sulfuric acid. Some of the most recognizable calcite from this district was found at the Shullsburg mine, where specimens consist of a single gray-white scalenohedron to 15 centimeters capped by a transparent rhombohedron to 10 centimeters on edge. The resulting scepterlike crystals (photo facing page) are highly prized by collectors. Golden-yellow, single scalenohedra to 24 centimeters that are twinned on {0001} and golden-yellow, complex low rhombohedra (to 10 centimeters) also occur at Shullsburg.

Illinois-Kentucky Fluorspar Districts

The MVT deposits located in southern Illinois and northwestern Kentucky have been important sources for fluorine and for minor amounts of lead, zinc and barium. Galena was reportedly being mined as early as 1819 near Cave in Rock, Illinois, and in 1875 former President Andrew Jackson began mining galena across the Ohio River in Kentucky (Lasmanis, 1989). The development of lead mining in the Rosiclare district led to the discovery of large vein systems that later became the nation's major source of fluorite.

The Bethel level of the Annabel Lee mine in Hardin County, Illinois produced this 6.5 cm calcite specimen. Ross Lillie collection; Jeff Scovil photo

Mineralization is localized in stratiform orebodies and in veins, or fissure fillings in faults that cut through the Mississippian-age limestones in the Rosiclare district, the Cave in Rock district and the adjacent Harris Creek district. The primary ore is *fluorspar*, the rock term for the mineral fluorite. In this deposit world-class fluorite crystals occur in association with beautiful calcite crystals. Calcite varies in size from microcrystals to individual doubly terminated crystals exceeding

Left: *Bill Severence owns this 3.2 cm high calcite with fluorite from the Denton mine in the Harris Creek district, Hardin County, Illinois. Photo Jeff Scovil*

Right: *Calcite (8cm high), twinned on {0001}, with celestine from the Bethel level of the Annabel Lee mine in Hardin County, Illinois; Ross Lillie specimen; Jeff Scovil photo*

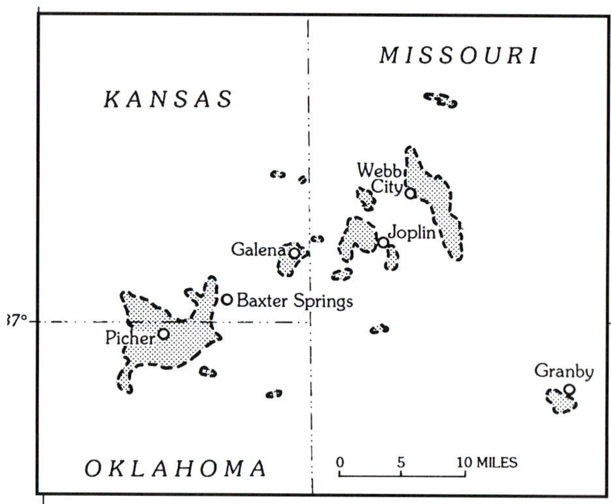

Overview of the Tri-State district (Lasmanis, 1989)

0.5 meter in length. Colors range from colorless and transparent to white, gray and amber. The dominant crystal forms are the scalenohedron and the rhombohedron. The classic calcites from the district are white to amber, extremely steep scalenohedrons (oppposite page) to 25 centimeters in length.

Prior to the opening of the Harris Creek district in the late 1970s, high-quality calcite specimens from the area were uncommon, if not rare. After its opening, however, no single orebody ever produced finer calcite specimens, in terms of variety and association, than the Denton mine (Lillie, 1986). At this mine, honey-colored calcites were found as lustrous dipyramids and with rhombohedral terminations on cubes of fluorite (above left). At the Annabel Lee mine, lustrous honey-colored scalenohedrons, twinned on {0001}, occur with crystals of celestine (above right). Alas in 1998, the mines were closed for economic reasons, with millions of tons of proven ore reserve. The shafts are capped and structures removed, and it is unlikely that these mines will reopen in my lifetime.

Tri-State District

This district is named for its location at the junction of three states. In Missouri, the district contains the Granby, Joplin and Webb City fields. The extensive Picher field is located in Oklahoma and Kansas. Mining began in 1848, and in the mid-1860s during the U.S. Civil War period, both Union and Confederate armies mined the deposits for lead. Zinc mining began at Granby in 1870, with production peaking in 1916 during World War I. In 1957 all operations were closed. The late Boodle Lane continued collecting specimens for years thereafter, until water reached the highest levels of the mine.

Nearly all the ore deposits were localized in limestones of Mississippian age that were heavily

Left: *An 11 cm wide calcite twin on $\{02\bar{2}1\}$ from the Tri State mine in Cardin, Oklahoma; Terry Huizing specimen and photo*

Right: *A 5.7 cm high calcite "phantom" on galena from the Sweetwater mine, in Reynolds County, Missouri; Terry Huizing specimen; Jeff Scovil photo*

faulted and altered by dolomitization and by the introduction of silica, which formed chert. The openings in which mineralization took place were enormous, and it was not uncommon to find simple calcite scalenohedra that were more than one meter long. Although nearly every museum and most private collections have specimens from this district, it is rare to find crystals that have not been affected by the corrosive fluids and, in later years, by the oxidation of sulfides. Possibly some of the most desirable calcite specimens from this district are the elongated steep rhombohedral crystals twinned on $\{02\bar{2}1\}$ from the Tri State mine and other mines in the Picher field (above left).

Viburnum Trend District

The largest MVT mining districts in the world are in southeastern Missouri, where the Old Lead Belt district near Bonne Terre was exploited from near-surface galena deposits as early as 1719. Underground workings were developed in 1864. By 1945 reserves in this mining district were seriously depleted; the last operation closed in 1972. Open space within the rock was limited, and few notable calcite specimens are preserved.

In 1955, to the west of the Old Lead Belt district and on the other side of the Ozark Dome, a blind drilling program in the Viburnum area resulted in the discovery of the northern extent of the Viburnum Trend district. Soon after, the southern end of the district was located more than 75 km away at Sweetwater. The district's huge ore deposits are located in Upper Cambrian dolomites and limestones. Mineralization is concentrated in reef complexes and collapse breccias; the best crystals are found in the open-space fillings in breccia zones. Unlike the Old Lead Belt deposits, every

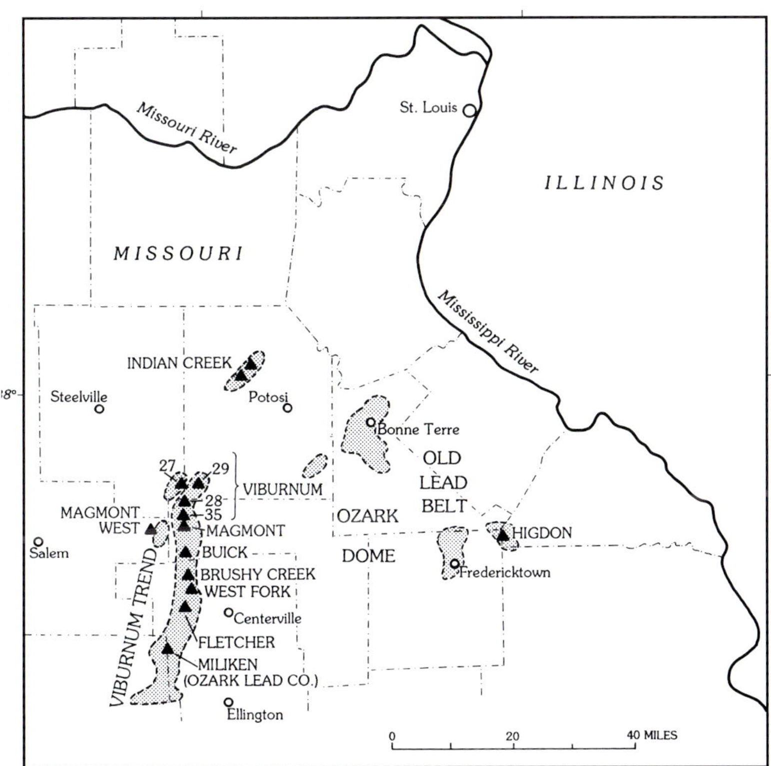

The Viburnum Trend district (Lasmanis, 1989)

Calcite "phantom" (13. cm long) on hematite from the Fletcher mine, in Reynolds County, Missouri Lawrence Nuelle specimen; Jeff Scovil photo

Calcite twin (6.9 cm high) on $\{0\bar{2}21\}$ from the Brushy Creek mine in Reynolds County, Missouri; Carolyn Manchester collection; Jeff Scovil photo

mine in the Viburnum Trend has encountered zones where minerals had space to crystallize (Lasmanis, 1989).

Viburnum Trend mines are known for beautiful calcite crystals that range to 20 centimeters in length. Colors vary from colorless and transparent, to white, golden-yellow and translucent gray. Phantoms of hematite, marcasite and pyrite dust the surface of crystals now overgrown with a clear generation of calcite (see photos this page and upper photo page 81). A variety of twins, particularly on $\{0001\}$, have been found at several mines. At the Brushy Creek mine, a small number of twins on $\{0\bar{2}21\}$ (below) were recovered. The dominant crystal forms are the scalenohedron, a steep rhombohedron and a variety of low rhombohedra. Minerals that are commonly associated with calcite include galena, dolomite, marcasite and chalcopyrite.

Central Tennessee District

This district is located east of Nashville in Smith County, Tennessee and extends to the northeast and southwest along the axis of the Cincinnati Arch. More than thirty shallow surface prospects have been known since the time of early settlers (Lasmanis, 1989). In 1967, a drilling program near Elmwood discovered the orebody in a dolomite of Middle Ordovician age.

Mineralization in this district is localized in collapse breccias related to the paleokarst surface of the dolomite. In open spaces within the breccia, the principle ore mineral sphalerite crystallizes with fluorite, barite and stunning transparent to golden-orange calcite scalenohedra. Calcite crystals to 50 centimeters in length are known, although those 20 to 25 centimeters long are more common. These superb crystals (facing page, lower right) are often twinned on $\{0001\}$ and could be among the best found in any mining district in the world.

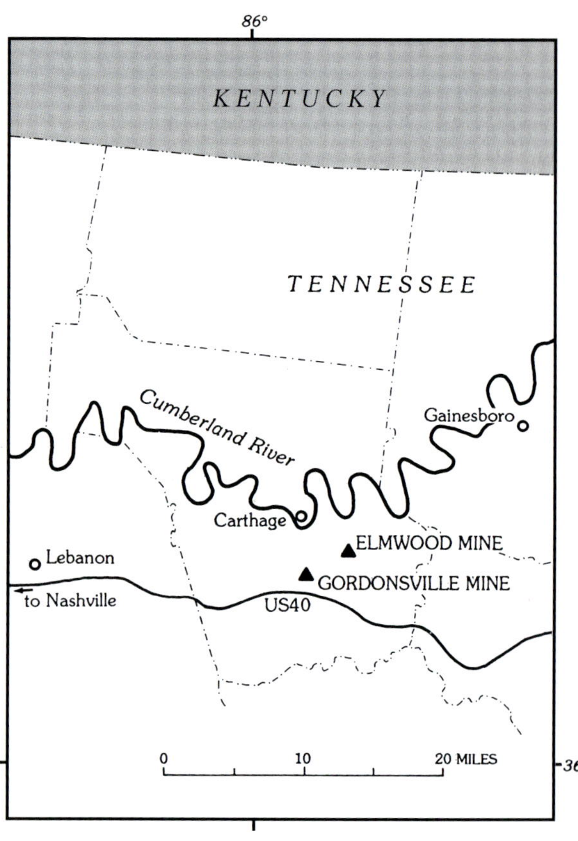

The central Tennessee district (Lasmanis, 1989)

An 18.8 cm wide calcite with marcasite from the Brushy Creek mine in Reynolds County, Missouri; Norma P. New specimen; Jeff Scovil photo

Limestone Quarries of the Midwest

As the basins of the midwestern United States filled with sediments, they also subsided. The sediments thickened and compacted, expelling fluids from the ancient saline seas upslope into older rocks along the flanks of the basin. In open spaces often provided by ancient reef structures, a simple suite of minerals crystallized in the carbonate rocks of Silurian and Devonian age. Unlike in the MVT deposits, galena is virtually absent; sphalerite, barite, fluorite and celestine are widespread and may be locally abundant. The carbonates, dolomite and calcite, are everywhere abundant within the limestones, and pyrite and marcasite are abundant within or adjacent to shales.

The large, colorful and sometimes twinned calcite crystals found in the reef structures of Indiana, on the eastern flank of the Illinois basin, are prized by collectors. Crystals may range to a meter in length, although it is more typical to find single crystals in the 10 to 15 centimeter range. Calcite varies from colorless and transparent to white and translucent and occurs in various shades of yellow, bright orange and golden-brown. All forms of calcite are found; common forms include the scalenohedron and rhombohedron. The prism and pinacoid are uncommon, but the dipyramid, a form that is elsewhere considered rare, is found with some abundance at many localities. Twins on $\{0001\}$ and $\{01\bar{1}2\}$ are common, but are rare on $\{02\bar{2}1\}$. Three Indiana occurrences, the Meshberger, Berry Materials and the Irving Materials quarries illustrate the diversity and quality of calcite found in midwestern limestones.

This 20.5 cm high calcite twin on {0001} from the Cumberland shaft in the famous Elmwood mine in Smith County, Tennessee is in the collection of Steve Neely.

Jeff Scovil photo

Limestone Quarries of the Midwest

Above: *The Meshberger quarry in Columbus, Indiana yielded this 16 cm calcite dipyramid. Terry Huizing specimen; Jeff Scovil photo*

Right: *A 3.5 cm high calcite twin on* $\{01\bar{1}2\}$ *from the Irving Materials quarry in Anderson, Indiana. Terry Huizing specimen and photo*

Below: *Calcite (10 cm high) with marcasite inclusions from the Berry Materials quarry in North Vernon, Indiana. Terry Huizing specimen and photo*

• At the **Meshberger quarry**, near Columbus in Bartholomew County, large crystals to 1 meter, but more commonly from 2 to 16 centimeters long, occur as lustrous golden-yellow scalenohedra or as dipyramids (left). Twins occur on $\{0001\}$, $\{01\bar{1}2\}$ and rarely on $\{02\bar{2}1\}$. Pyrite and fluorite crystals and natural asphaltum are associated with the calcite.

• The **Berry Materials quarry** in North Vernon, Jennings County has yielded distinctive, doubly terminated crystals to 8 centimeters that are

nearly equal in all dimensions. These consist of a busy combination of dipyramid, steep rhombohedron, prism and scalenohedron terminated by a prominent large, low rhombohedron (left). Some calcite crystals may be included with crystals of marcasite. Crystals of iron-rich sphalerite are locally abundant.

• The **Irving Materials quarry** in Anderson, Madison County produces translucent, yellow to orange calcite crystals to 15 centimeters, though they are more commonly from 3 to 5 centimeters in length. Crystals consist of equally developed dipyramid and scalenohedron, terminated by a low rhombohedron. These are twinned on $\{01\bar{1}2\}$ (above) and rarely on $\{02\bar{2}1\}$. Crystals are loosely attached to the limestone matrix and may be easily plucked from the vugs in which they formed.

Michigan Calcite and Copper

by ***Stanley Dyl*** *from Houghton, Michigan and* ***Terry Huizing*** *from Cincinnati, Ohio*

Approximately 1.2 billion years ago vast quantities of flood basalts were extruded into the mid-continent of North America. This ancient rift system is presently exposed only in the western portion of the Lake Superior district in Michigan. Mineralizing solutions consisting of slightly alkaline fluids later percolated upward following fault and fracture zones and at temperatures ranging from 180° to 300° C, deposited copper and native silver in open space within the basalts.

Spectacular calcite crystals have been produced from various mines throughout the district. Crystals range from microscopic to 18 centimeters in length. They range from clear and colorless to pale yellow, gray, brown, orange, black and red to pure white (Wilson, 1992). Twins on {0001} are common, as are numerous complex habits created by hundreds of uncommon crystal forms. By far the most impressive of these are transparent scalenohedral crystals with bright copper phantoms inside.

Left: *A 3.1 cm wide calcite on native copper from Kearsarge, Houghton County, Michigan. Dawn Minette collection; Jeff Scovil photo*

Below: *The Quincy mine area in 1915: view from the rockhouse of the No.2 shaft looking north toward shaft No.6 Photo courtesy Michigan Technological University*

Lustrous, sharp calcite crystals with native copper inclusions from Michigan's Keweenaw Peninsula on Lake Superior are highly desirable, classic mineral specimens. While most of Michigan's native copper lodes produced specimens of copper-in-calcite, the premier source for this unique association is arguably the Pewabic Amygdaloid Lode in Houghton County. The Franklin, Pewabic, Quincy and Mesnard mines all worked the Pewabic Lode and have produced the greatest abundance of Michigan's fine to world-class specimens of crystallized calcite with copper inclusions.

Underground workings at the Pewabic Lode have long been idle, but its mineral heritage lives on, accessible to visitors of the Keweenaw National Historical Park – the only historical park in the National Park system devoted to the story of underground mining. In addition, the century-old A. E. Seaman Mineral Museum is planning to relocate to a new facility on the site. The museum will occupy a space based on two historic Quincy mine buildings, creating an impressive mineral collection at the mine.

The A. E. Seaman Mineral Museum has one of the top mineral collections in North America and has the world's best collection of Michigan minerals, especially from its copper and iron mining districts. The museum's premier suite of Michigan copper-in-calcite specimens will be displayed in the new museum facility, projected to be completed in 2008. The suite includes the almost perfectly-proportioned 8.5 centimeters tall *Rocket* from the Franklin mine as well as a spectacular 19.5 centimeters wide Quincy mine copper-in-calcite specimen with 9 centimeters crystals from the Captain Thomas Whittle Collection. These calcite specimens have long been considered some of the world's best.

Calcite Crystals in Amethyst Vugs

Collector ***Reinhard Balzer*** *from Marburg, Germany visits a famous amethyst locality that happens to boast a surprising number of calcite habits.*

The mining town of Irai, in the northwestern corner of the Brazilian state of Rio Grande do Sul, is well known for producing wonderful amethyst geode specimens. In the not so distant past, however, miners blasted the gem-rich cavities out of the local volcanic rocks, smashed them and collected the violet crystals that bedecked the walls, without regard for the concerns of specimen collectors. Miners were commissioned only to produce as much amethyst gem rough as possible.

In the early 1990s, however, things began to change. More attention was paid to working large geodes, that contain beautifully formed amethyst druses, out of the host rock in one piece. These geodes are now widely available on the mineral market where they often bring a higher price than they would have brought as gem rough. When such a vug includes a choice group of calcite crystals perched on amethyst, it is even more valuable.

The German Influence

Rio Grande do Sul is the most southern state in the Republic of Brazil. It shares its northern border with the state of Santa Catarina, its southern border with Uruguay and its western border with Argentina; in the east, it borders on the Atlantic Ocean. Porto Alegre is its capital, with Rio Grande its most important port. The climate of the state is temperate to subtropical, but it can get quite cold in the winter months, and it even snows in the mountains from time to time. Geologically, Rio Grande do Sul is part of the Parana Basin, which stretches north to south from Parana, through Santa Catarina and into Uruguay.

The author would like to express his thanks to Orlando Klein of Soledade, Brazil who helped him to gain access to the mines and who provided much of the information found in this report.

Because its landscape is in many ways similar to that of Germany's central mountains, it is not surprising that Rio Grande do Sul was one of the most significant emigration destinations for 18th and 19th century Germans. A large portion of these German emigrants came from the Hunsrück, Eifel and Pfalz regions. The dialects of the homeland are still spoken in Rio Grande do Sul, in spite of their prohibited use during WW II.

In the early 19th century, abject poverty reigned in many parts of Germany, especially in the west, driving masses of people into the hands of promoters who organized emigrations to destinations around the world. Emigrations often ended in misery and disappointment, as the flights were often badly organized and did not lead to the promised ends.

Poverty and romantic preconceptions of distant paradise, however, were not the only attractions of emigration. In his 1989 book *Auswanderung von Hunsrück nach Brasilien* (Emigration from Hunsrück to Brazil), August Meter writes, "Decades of war, robbery and plunder by French armies, exploitation of the country by duties and taxes, as well as forced recruitment of young men into the Napoleonic army on its way to Russia, brought unspeakable harm to Germany, and the Hunsrück, Pfalz and Eifel in particular. The causes of the emigration from this region have to be understood in light of the economic and political conditions after the Napoleonic wars and the French revolution."

The Cutters Emigrate

At the beginning of the 19th century, lapidary centers Idar and Oberstein as well as villages that surrounded them had exhausted the local sources of cutting rough. Some families emigrated especially to the state of Rio Grande do Sul in Brazil to earn their living in a new homeland. Most cutters were farmers in the old country, and they found the climate and soil conditions in southern Brazil ideal for planting. A thriving agriculture quickly developed, and thanks to these industrious people, Rio Grande do Sul is today the breadbasket of Brazil.

In 1827, the son of an emigrant cutter from the Hunsrück found, quite by accident, a familiar stone in a riverbed: a piece of fine quality agate, a gift from heaven. By 1834 the first shipment of agate was carried by ship across the Atlantic, by barge up the Rhein and by oxcart into the Hunsrück.

The lapidary industry in Germany was saved, and the following decades saw rapid growth. Systematic prospecting began immediately in Rio Grande do Sul, and large reserves of amethyst, citrine, smoky quartz and gemstones were uncovered in addition to the coveted agate.

Irai, Rio Grande do Sul, Brazil

Landscape Similar to that of Idar-Oberstein

View from a calcite-amethyst mine of the rolling hills near Irai

Right: *A collector's dream, the quarry walls are peppered with vugs.*

Left: *Big finds can come from small quarries.*

Photos by Reinhard Balzer

The Creative Power of Volcanoes

Rio Grande do Sul is blanketed by old lava flows. Sometimes numerous individual flows lie on top of one another, separated by intercalated sedimentary deposits. These *lava series* have varying thicknesses that can reach up to 1 kilometer and are roughly 90 to 100 million years old. Cavities left by the outgassing of the lava were lined and filled with minerals, principally quartz in the form of agate and amethyst. Calcite and gypsum crystals are more recent growths.

Vugs that were completely filled with agate are known as *amygdules*. Those that are lined with crystals are referred to as *druses*. Cavities can reach 4 meters in length and more than 1 meter in diameter. The exterior shapes of complete vugs that are extracted from the host rock reflect the interplay between the escaping gasses that formed them and the flow of the lava that surrounds them.

If the uppermost layer of lava is near the surface of the ground, the deposit is worked through an open pit, an option that is easier and less dangerous than mining underground. Where underground workings are warranted, adits are mechanically dug along the productive layers of rock. Homemade black powder serves as the explosive, as commercially available dynamite is far too strong and would destroy the delicate crystals, especially the calcite.

When a vug is encountered and partially exposed, a peephole is carefully driven into it. Probing it with a light, miners assess the vug deciding whether to extract it. If the quality of the crystals does not warrant extraction of the vug as a whole, it is broken up for gem rough.

Buried Treasure in Irai

Upper left: *Celadonite is included in the crystals of this 13 cm wide calcite group. Collection Terry Huizing; photo Jeff Scovil*

Left: *After carefully making a small opening, a miner checks the vug to discover whether it is durable enough to endure a blast. Photo Reinhard Balzer*

Above: *This 8.1 cm wide calcite twin on* $\{10\bar{1}1\}$ *is a part of Victor Yount's extensive calcite collection. Photo Jeff Scovil*

From Vugs in Volcanic Rocks

Dream Specimen from Irai: *This 18 x 9 cm "floater" consists entirely of calcite crystals. Photo by Hartmut Meyer*

The Calcite

More than 350 distinct habits of calcite have come from the amethyst vugs of Rio Grande do Sul (pers. comm., Victor Yount, 2003). The largest and most beautiful calcite is found around the northern towns of Frederico Westphalen, Plan Alto and Irai, where there are innumerable outcrops and dozens of productive localities.

The underground workings in the area of Irai produce the most notable crystals. These are typically elongated scalenohedral crystals that often taper to a point at both ends. Numerous small points sticking out of the crystal faces sometimes give the impression that big crystals are built from many smaller ones. The overall crystals can be as long as 20 centimeters and often stand in bizarre groups aggregated in the hollow cavity of the amethyst geode. The groups are sometimes chaotically and sometimes radially organized. Calcite also occurs in compact, rounded habits. These too are often formed from numerous tiny crystals. Although the scalenohedron form predominates, calcite of a prismatic habit and terminated by the pinacoid, occurs. Remnants of these crystals are sometimes preserved as hollow, quartz covered, encrustation pseudomorphs.

Calcite crystals from Irai are generally colorless and milky, though they sometimes have a yellow tinge. Their faces often have a fine, silky luster. While amethyst is constantly being mined at Rio Grande do Sul, good calcite specimens on amethyst are found more sporadically, and really super finds turn up only every couple years.

The green mineral that is omnipresent on the outer surface of the vugs is *celadonite*, a dioctahedral potassium, iron, magnesium mica. Celadonite is also the included mineral found as green *phantoms* within calcite crystals (Carl Francis, pers. com.). A point of caution, buyers

Left: Calcite Corsage

This 10.8 cm high calcite arrangement was being offered for sale in 2002 by Fine Minerals International. Jeff Scovil photo

of the large crystal-lined vugs should be aware that it now seems to be common practice to apply a liberal coating of celadonite-color paint over the outer surface to conceal areas chipped or damaged in the extraction process.

Calcite is not the only byproduct of amethyst mining. Magnificent gypsum aggregates, though rarer than calcite specimens, are also found in the vugs. Notably, not every vug contains the classic violet amethyst crystals; smoky brown quartz and natural citrine crystals also line the cavities, but these are great rarities. Amethyst from these mines, both individual crystals and whole druses, are often burned in ovens until they are yellow, at which point they are sold as *citrine*; however, this process produces a characteristic honey-yellow color that the trained eye can immediately distinguish from natural citrine.

Of course, calcite in amethyst geodes is found in localities outside of Rio Grande do Sul. In principle, specimens of this nature can occur in any deposit that is geologically similar to those mined in Brazil. The occurrences at Fischbachtal and other localities near

A 7 cm high yellow calcite crystal sits in an amethyst vug from Irai. This specimen, from the collection Reinhard Balzer, is on display at the Mineralogical Museum in Marburg, Germany. Hartmut Meyer photo

Idar-Oberstein have produced similar specimens that have been sought by collectors for years. The prizes from these Old World localities include complex calcite crystals in amethyst and smoky quartz druses. Respectable calcite specimens have also turned up in the spherical amethyst-containing geodes from Chihuahua, Mexico.

The expansive amethyst belt in Uruguay (Catalan district and others) from time to time produces, among other things, the famous *skunks*, large white to yellow calcite crystals to 13 centimeters with overgrowths of deep purple amethyst and zoned inclusions of a black mineral, coupled with oriented overgrowths of small, steep, white, second-generation calcite crystals. The amethyst overgrowths are arranged like stripes a couple of centimeters wide running from the termination down the center of the steep rhombohedron faces.

Massachusetts artist Frederick C. Wilda's watercolor rendering of a dream amethyst on calcite "skunk." The specimen that inspired this portrait was found in town of Artigas in the Uruguayan province of Artigas and is in the collection of long time dealer/ collector Rock Currier. A photograph of the actual specimen graced the March/April 1997 issue of Rocks & Minerals.

Photo Gallery

Above:
3 cm wide calcite and dioptase
Tsumeb, Namibia
Roland Sherman collection
Jeff Scovil photo

Above:
6.5 cm high stalactitic calcite
Carter County, Montana
Michael Cooper specimen and photo

With innumerable worldwide localities for calcite, it is literally impossible to cover them all in a single volume or even review all of the significant deposits.

Tsumeb

Tsumeb, Namibia is a perfect example. Its copper-lead-zinc orebody carries more than 250 different mineral species. Tsumeb is among the most species-rich localities on Earth. Calcite is the most common mineral found at Tsumeb. It occurs in association with nearly all of the other minerals. At Tsumeb, the rhombohedron is the most common calcite habit, and associations with and inclusions of various other mineral species have provided for wonderfully colorful calcite crystals.

Mexico

The enormous gold-silver province of central Mexico has provided collectors with exceptional calcite specimens. Stretching from the state of Chihuahua in the northwest through the state of Guerrero in the southeast, 1,000 km across the Sierra Madre Occidental, these ore deposits host the largest silver enrichment in the world.

The silver mines are active today and a stream of calcite specimens continue to trickle out of central Mexico.

And All the Rest...

Even relatively obscure localities have at times produced world-class calcite specimens, as have many of the deposits worldwide that are known for other minerals.
Though we are unable to share them all with you, we wanted to show at least a smattering of photographs of calcite from other localities around the world. **Enjoy!**

Above:
9.5 cm high calcite
Woodlawn quarry, McDowell County, North Carolina
private collection

Left:
12.5 cm high calcite, hematite and duftite Tsumeb, Namibia
Marshall Sussman collection

Facing Page:
4 cm high calcite with chalcotrichite inclusions on copper
Onganja, Namibia
Carolyn Manchester collection

All photos Jeff Scovil

Upper right:
8.1 cm wide calcite and quartz; Juchem quarry, Idar-Oberstein, Germany; collection Hoffman-Rothe
Jeff Scovil photo

Upper left:
11.5 cm high crystal group from Rudnui, Kazakhstan; collection François Lietard
Michael Cooper photo

Middle right:
13.6 cm high specimen of calcite
Malmberget mine, Gällivare, Lappland, Sweden
Peter Lyckberg collection; Jeff Scovil photo

Lower left:
4.7 cm wide specimen of calcite and malachite
130 m level, Red Dome Gold mine, Cillagoe, N. Queensland, Australia
Evan Jones collection; Jeff Scovil photo

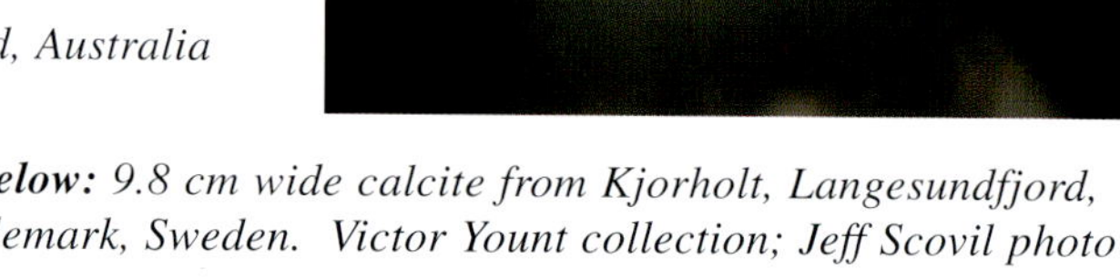

Below: *9.8 cm wide calcite from Kjorholt, Langesundfjord, Telemark, Sweden. Victor Yount collection; Jeff Scovil photo*

Mexico

Right:
2.4 cm wide color-zoned calcite
Guanajuato, Guanajuato, Mexico
Judy Megerle collection
photo Jeff Scovil

Center:
A statue adorns a public square in Charcas, Mexico celebrating the city's mining heritage
Günther Neumeier photo

Center right:
8 cm high calcite crystal aggregate
Charcas, San Luis Potosi, Mexico
collection Fred Pough
Jeff Scovil photo

Lower right:
9.4 cm wide "manganoan" calcite
Aranzazu mine, Concepción del Oro, Zacatecas
collection Bill Larson
photo Jeff Scovil

Below:
8.5 cm high calcite group from San Martin, Zacatecas, Mexico
Francis Benjamin collection
photo Jeff Scovil

Limestone Caves: Wonderlands of Calcite and Aragonite

***Sarah Bronko**, aspiring scientist and student at Westover School in Middlebury, Connecticut on the marvels of limestone caves*

From stalactites and stalagmites to soda straws and flowstone, the architecture of a limestone cave is the result of the dissolution and precipitation of calcium carbonate. Known as *speleothems*, derived from the Greek words *spelaion* (cave) and *thema* (deposit), cave formations are dependent on the mobility of calcite, the primary mineral component of limestone.

Calcite's readiness to dissolve has provided the medium for the speleothems that are often found in limestone caves. As water leaches through the organic matter in soil, it picks up relatively high concentrations of carbon dioxide, lowering its pH. This slightly acidic water seeps through cracks in limestone, dissolving calcite and carrying it along with the water and carbon dioxide. Once the solution emerges into an open cave, the carbon dioxide diffuses into the surrounding air, which has a lower concentration of carbon dioxide; this causes calcium carbonate to be released from the water, and it forms a crystalline deposit.

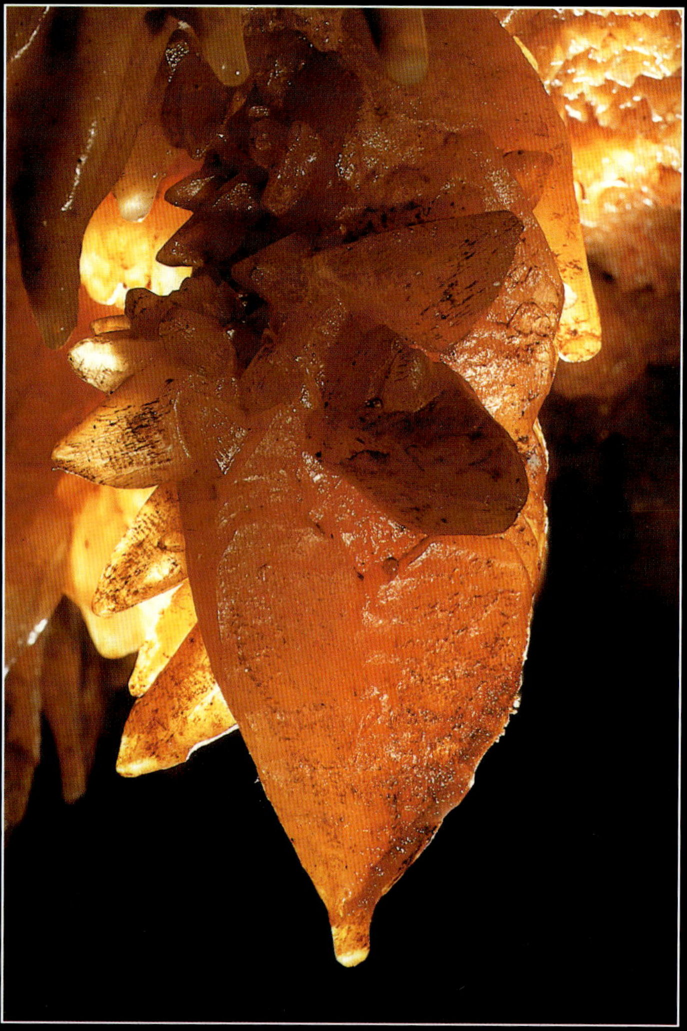

***One Crystal:** A 25 cm long stalactite is a real rarity. This calcite crystal is still forming in Algar Gantua, Portugal. Photo Alain Martaud*

An Array of Formations

The nature of the flows from which calcium carbonate precipitates determines the size and shape of the deposits. The successive dripping of calcite-saturated fluid from the cave ceiling often forms hollow tubes, or **soda straws**, one of a number of standard calcite formations. The soda straws also fill up with deposited calcite, but they fill more slowly than they form. Once a soda straw is full, fluid streams down the outside of the formation, thickening it into a **stalactite**. Conversely, **stalagmites** are produced when calcite-saturated water drips to the ground where it forms a dense deposit.

Helictites are contorted speleothems that twist in any direction. They grow on ceilings, walls or on other speleothems; they typically occur with soda straws. These curved and angular, lateral projections form when saturated water, under hydrostatic pressure, passes through tiny central canals and is forced through a capillary channel and out a small pore. Fluid can be forced through pores in any direction, resulting in depositions that seem to defy gravity.

Flowstone, a large, cascading speleothem, is the product of a steady stream of water that continuously evaporates while depositing calcite into long, often undulating sheet-like formations. Conversely, thin speleothems known as ***cave veils*** are formed when a thin stream of water drizzles along a slanted ceiling to create a delicate calcite drapery.

Cave pearls are small speleothems that evolve in shallow cave pools as layers of precipitated calcite that often form around a nucleus of sand or broken fragments of other caves formations. Cave pearls form singly or together in seeming nests. While cave pearls are often spherical or elliptical, they occasionally take the shape of cubes!

Cave rafts, also referred to as *cave ice*, form on the surface of still, supersaturated pools, nucleating around dust and particles on the water's surface. Calcite attaches to the rafts in circular patterns and is supported by the surface tension of the water, forming an intricate surface crust. If the water remains still and the growth of the delicate raft is not

A Soda Straw Forest

These straws are 2-3 meters in average length with some over 5 meters. All are active, hollow and incredibly delicate. The person in the image is well behind the straws.

Routes of travel through this secret French grotto are carefully chosen in order to avoid the deposits and prevent accidental damage to the formations.

The photographs for this essay were taken (unless otherwise noted) by Kevin Downey of Northampton, Massachusetts. Kevin is a commercial photographer and avid caver. He has been exploring and photographing caves for more than 35 years. Kevin has taken his camera on more than 2,200 trips underground in 42 countries.

is the primary calcium carbonate mineral deposited (Hill, 1986). Because Mg^{2+} is generally excluded from the structures of both calcite and aragonite, it is thought that a high concentration of the magnesium ion poisons the crystal growth of calcite, allowing aragonite to precipitate in arrays of lacy, intricate **frostwork** and **cave coral**.

The main difference between calcite speleothems and *normal* calcite crystals is that speleothems are usually made up of many tiny crystals of calcite, not just one or a few. This is a result of more rapid deposition of calcite in caves than is typical in environments noted for well-formed calcite crystals. Speleothem habits are, thus, not significantly influenced by the forms that govern the growth of

Calcite Cups
These hollow structures reach over 13 cm high. Their triangular cross-section suggests that each of them is a single crystal. Growth occurs primarily on the rims of the cups, which reach the surface of the pool in which they are growing. These forms, located in a cave in northern Cuba are both rare and delicate.

interrupted, the weight of the raft itself will eventually causes it to sink. **Cave shelves** extend from the edge of a cave pool, forming a crust over its surface. As the water level changes, new shelves may form above and/or below existing shelves. Often sturdier than cave rafts, these series of shelves can remain intact even after the water has entirely receded from the cave.

Many speleothems are colored by impurities such as metal ions in the ground soil, through which the solutions first flow. Each impurity causes a different hue in calcite. For instance, a high concentration of iron in the soil above the cave may result in a vibrant red and orange speleothem, blue and green are often attributed to copper, black and gray to manganese and muddy-brown, tan and rose-red to clays, silt and mud. As with crystallized minerals, the exact cause of color in speleothems is not always well understood.

Free-Form and Rapidly Growing

Two hundred eighty mineral species are known to occur in caves, including all three of the polymorphs of calcium carbonate. Of the polymorphs, vaterite is the least common because cave temperatures are generally too low for it to form. After calcite, aragonite is the most commonly found. While cave temperatures and pressures favor the growth of calcite, the presence of the magnesium ion influences aragonite deposition. In fact, when the Mg^{2+} to Ca^{2+} ratio reaches about 3:1, aragonite

The tips of these Lechuguilla cave stalactites are coated with secondary pool deposits of mammillary calcite. Roughly 4.5 cm in diameter, these balls show distinct water level fluctuations.
Photo Kevin Downey and Urs Widmer

Subterranean Oasis

A deep pool surrounded by calcite "shelves" and aragonite "coral." This site is known as the "Lake of the Blue Giants" and is in Lechuguilla cave in Carlsbad, New Mexico. Lechuguilla is the deepest cave in the United States. This pool reaches depths of more than 30 meters.

Photo Kevin Downey and Urs Widmer

Above: Dinner Anyone?

Traditional, spherical calcite cave pearls formed between more exotic "ravioli" in a cave in southern France. Photo Kevin Downy

Left: *A 13 cm high spray of pure white aragonite formed on a pool deposit of peach calcite in the Lechuguilla cave, New Mexico*
Photo Kevin Downey and Urs Widmer

For more information on the study and conservation of caves, contact the National Speleologic Society, 2813 Cave Avenue, Huntsville, AL 35810, USA, telephone 256-852-1300, www.caves.org.

Rafts, Pearls, Lamps and Lily Pads

Cave Ice

Ice-like calcite deposits form on the surface of a Lechuguilla cave pool basin. If these overgrowths remain undisturbed, they may eventually cover the surface. These deposits frequently include small rafts of floating calcite that can act as nuclei for the further precipitation of the mineral.

Photo Downey and Widmer

Colorful Copper

***Above**: Electric blue, acicular aragonite is unique to this secret site in France; the color is the result of copper impurities.* ***Right:** This green stalactite with an aragonite overgrowth derives its color from trace elements, primarily nickel. Copper is not an active ion at this deposit in southern France.*

individual calcite crystals. They are, instead, free-form structures created in layers and are defined by the patterns of the thin films of the water that deliver calcite for their growth (Hill, 1986). In some situations fairly well-formed crystals are found as part of speleothems. Not uncommonly, slower-growing stalactites have tips that are composed of single crystals that show the standard cleavage of calcite and crude to well developed crystal faces.

Protected By Law and Conscience

In an attempt to protect natural cave resources and formations, the United States legislature passed the Federal Cave Resources Protection Act of 1989. This bill guards caves in national parks, forests and other land administered by the Departments of Agriculture and the Interior. By 2001 twenty-six states had also enacted cave protection laws in an attempt to shield their caves from the inevitable destruction caused by the insatiable interest of cavers and tourists. Laws protecting caves have also been passed internationally and conservation associations have arisen in several international communities.

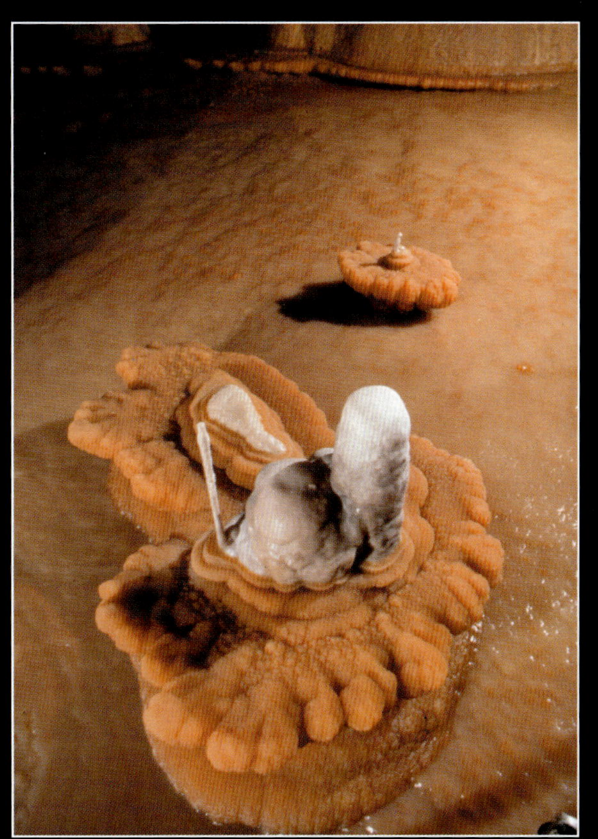

Lamps and Lily Pads

In a cave in northern Cuba, calcite has crystallized into transparent "lamps" that hang from stalactites in a former pool zone. The largest of these structures is more than 1 meter long.

Massive calcite lines a shallow pool in the "Oasis Pool Room" in the Lechuguilla cave, New Mexico. The pools gave rise to lily pads on which secondary stalagmites have formed.

Carlsbad Curtain

A calcite drapery hangs across a small canyon passage in the left tunnel region of Carlsbad Caverns, New Mexico. Photo Kevin Downey and Urs Widmer

Upper left:
Filamentary helictites radiate from a stalactite to form a hanging bush that adorns Caverns of Sonora on the Edwards Plateau in Texas.
Photo Downey and Widmer

Upper right:
In a deposit in Cuba, calcite crystals developed unusual secondary helictites on their terminations.

Left:
Monocrystalline and helictitic calcite fringes a stalactite and forms a small island in a cave pool in Transylvania, Romania

Antlers and Showering Water

Helictites branch like antlers across a section of passage in a Colorado caveceiling. Some of the forms are "beaded," exhibiting bulbous protrusions that have a fibrous structure running parallel to the direction of growth. Photo Kevin Downey and Urs Widmer

Right: *Helictites commonly branch out in response to slight changes in local conditions. The darker parts of this gem helectite, found in l'Hérault, France, consist of fibrous aragonite, the white tips are pure calcite. Photo Kevin Downey*

Far right: *A crystal group forms in a water droplet in Souterroscope des Ardoisienes, Calvados, France Photo Alain Martaud*

Waterworks

This active cave "showerhead" was found raining in a cave in Puerto Rico. This form is simultaneously precipitated and dissolved.

Photo Kevin Downey

Cleaning Calcite and Aragonite

Chemist and mineral cleaning specialist ***Rudolf Duthaler*** *from Basel, Switzerland with a recipe for cleaning thin rust stains from calcite and aragonite specimens*

Cleaning mineral specimens that include calcite or aragonite is a rather delicate task given that calcium carbonate crystals have a low hardness and are moderately soluble in water. At neutral pH, the solubility of calcium carbonate ranges from 14 to 15 mg per liter in cold water to 18 to 19 mg per liter in hot water. This is actually pretty low: a few sand-size grains per liter of water. When, however, these few grains come from the faces of your favorite calcite crystal, they can make the difference between a glowing gem and a dull disappointment.

The solubility of calcium carbonate increases sharply when the pH of the water is lowered into the acidic range or when chelating agents, found in some common detergents, are incorporated into the cleaning solution. Even very weak acids such as carbonic acid (e.g., club soda, seltzer and rainwater) increase the solubility of both calcite and aragonite. While strong acids completely dissolve calcite or aragonite, it is possible to clean these minerals without damaging them. Specimens can be washed with water and iron stains removed with sodium dithionite.

Mechanical Cleaning

Plant roots and common dirt such as dust, earth and clay are best removed from specimens by washing them with cold tap water and mild liquid dish soap. Stronger detergents such as dishwasher detergent or those used for lab glassware are not recommended as they generally contain chelating agents that etch calcite and aragonite, rendering shiny crystal faces dull. Hard tap water, already partially saturated with calcium ions, is preferred over distilled, deionized or even rainwater, which is often quite acidic both from natural carbon dioxide and from atmospheric pollutants such as sulfur and nitrogen oxides. Prolonged soaking in water should be avoided; in fact, all procedures for cleaning calcium carbonate should be done expeditiously.

Because hard dirt particles may be present and are likely to scratch the surface of a crystal, cleaning calcite or aragonite specimens with any brush, no matter how soft, should be avoided; thus, the best mechanical procedure for removing dirt from calcite or aragonite specimens is a pressurized stream of water. A shower spray, or better yet a dental *water pick*, provides a pulsating jet with an adjustable intensity and focus. Of course, cleaning specimens that are heavily contaminated with dirt can be quite a messy process. Protective eyeglasses are recommended, and the basement or garage may be a better workspace than the kitchen or bath. Cleaning with a water jet should be done over a plastic container to trap dirt and keep to household plumbing from clogging.

This and many other methods for cleaning minerals are published in Duthaler and Weiss' German-language book Mineralien Reinigen und Aufbewahren.

Cleaning with Sodium Dithionite

Rust is a general term that describes various hydrated ferric (Fe^{3+}) oxides and hydroxides that are prevalent and are often a major focus when cleaning minerals. Direct rust removal is only possible with rather strong acids, typically citric, oxalic or hydrochloric. Strong chelating agents such as EDTA (ethylenediamine-tetraacetic *acid*) or NTA (nitrilotriacetic *acid*) also dissolve ferric iron; however, these reactions are much too slow for practical mineral cleaning. As calcite and aragonite do not withstand acid treatments, cleaning rust from these minerals is usually considered impossible.

This is not necessarily the case, however, as sodium dithionite reduces ferric iron to more soluble ferrous iron (see Equation 1). Sodium dithionite has the chemical formula $Na_2S_2O_4$. The 10th edition of the *Merck Index* notes that the term *sodium dithionite* is commercially applied to a number of different products. **It is important to identify the substance being purchased by its chemical formula rather than by its name.**

Rust removal using sodium dithionite can occur in a pH neutral environment and would thus be the perfect method were it not for the fact that the rather unstable dithionite reaction always produces acidic byproducts that can etch calcite (see Equation 2).

$$2Fe^{3+}(OH)_3 + Na_2S_2O_4 \longrightarrow$$
$$Fe^{2+}(OH)_2 + Fe^{2+}SO_3 + Na_2SO_3 + 2H_2O \quad (1)$$
$$2Na_2S_2O_4 + H_2O \rightarrow Na_2S_2O_3 + 2NaHSO_3 \quad (2)$$

For this reason a mild alkali, usually sodium bicarbonate ($NaHCO_3$; baking soda), is always added to dithionite solutions to keep the pH from becoming acidic. A chelating agent (*e.g.* sodium citrate or EDTA) is often added to keep the iron in solution. While this technique works very well for many minerals, chelating agents also have an affinity for calcium ions and tend to dissolve them; thus, chelating agents must not be used when treating specimens with calcite or aragonite.

Modified, the sodium dithionite cleaning procedure has been successfully used to remove thin iron hydroxide coatings from calcite or aragonite specimens. To use the procedure effectively it is important to avoid chelating agents such as sodium citrate, reduce the concentration of sodium dithionite and sodium bicarbonate to about 1/10 the amount generally used for cleaning other minerals, and to monitor carefully the pH of the solution, adjusting it to neutral or slightly alkaline as necessary.

Procedure for Dithionite Treatment

Safety Precautions: Sodium dithionite has an unpleasant smell, like that of rotten radishes; thus, work should be done either outside or in a well ventilated place. Gloves, protective glasses and a dust mask are highly recommended. In high humidity, the spontaneous ignition of dithionite dust is a risk. In solution, sodium dithionite forms sulfur dioxide fumes, especially at acidic pH or in high concentrations; this toxic gas can provoke severe attacks of asthma.

The specimens to be treated should first be mechanically cleaned, then placed in a plastic container that can be closed with a cover. Estimate the amount of solution that will be needed to cover the pieces in the container. For each estimated liter of cleaning solution, place, in a second similar-sized container, approximately 2 grams each of granular sodium dithionite and granular sodium bicarbonate (standard baking soda will do). Add tap water and stir to ensure complete dissolution of the chemicals. This cleaning solution should immediately be added to the container with the mineral specimens. Adjust the final volume of solution to cover all of the specimens in the container by rinsing the mixing container with water then adding more water if necessary. To contain the unpleasant and toxic fumes, cover the container for the cleaning period, which should not exceed three hours.

Remember that a byproduct of the intended reaction is acidic sodium bisulfite (see Equation 2); thus, the cleaning progress should be monitored at least hourly using commercially available pH paper, to be sure that the pH of the solution remains in the alkaline range. To avoid damage to the acid sensitive specimens, more baking soda should be added if the pH falls below 7. A white, cloudy solution indicates that elemental sulfur is being precipitated, a decomposition process observed at low (acidic) pH and high dithionite concentrations. If the specimens, or worse the solution, become black, all of the samples should be immediately removed from the solution and flushed thoroughly with tap water. The black color is either due to mixed ferric/ferrous iron complexes (that often form in limonite crusts) or to the presence of minerals that are incompatible with dithionite (see below).

After soaking for from 30 minutes to a maximum of 3 hours, the mineral specimens should be removed from the cleaning bath (wear gloves!) and flushed with tap water mixed with mild dish soap. Additional treatment with a water jet can aid in the removal of sulfur precipitates. In a few days, the air will oxidize to their original brown any dark areas that remain on the specimens after treatment.

The oxidation process can be accelerated with an oxidant such as hydrogen peroxide (less than 100 ml of 3 percent H_2O_2 per liter of water) or a little bleach (sodium hypochlorite, NaOCl). The standard concentration of hydrogen peroxide available in drug stores is 3 percent. Commercially available hydrogen peroxide solutions are slightly acidic; thus, bleach is the preferred oxidizer. Cleaning can be repeated if stains reappear after drying.

Dithionite treatment is recommended for removing rust from many minerals in addition to calcite and aragonite. There are, however, quite a number of minerals that can be damaged by the reductive power of dithionite. These are essentially the highly colored minerals found in the oxidation zones of ore deposits. Among them are many Cu^{2+} minerals (*e.g.*, malachite, azurite and chrysocolla) as well as minerals containing cobalt, vanadium or lead. In addition, silver and mercury minerals (*e.g.*, chlorargyrite) are sensitive to thiosulfate, a decomposition product always present when working with dithionite (see Equation 2).

Collector and cartoonist Marcel Vanek's unique approach to large cleaning projects

If there is any doubt as to whether a sample will tolerate dithionite treatment, test the treatment on a less valuable specimen. It is always advisable to consider whether the aesthetic outcome of cleaning is justified. Removing an iron stain from one crystal in a group may reduce the overall quality of the specimen if attractive color contrasts are lost. In addition, any cleaning process involving chemicals can readily etch the more lustrous surfaces of calcite crystals. For such specimens, cleaning processes should be kept quite short. A dithionite treatment is almost fully effective in 10 to 30 minutes. When you find yourself wondering whether the specimens should be in the bath a bit longer, the answer is probably, "No."

We are grateful to George Megerle for his assistance with this article.

A Skeleton Made of Calcite Crystals

Rupert Hochleitner *and* ***Pete Richards*** *report on the calcite crystalline makeup of a unique phylum of marine animals.*

Echinoderms are slow moving or attached organisms found in a variety of saltwater environments. Many echinoderms have spines or other appendages, for which the group is named (*echinoderm* means *spiny skin*). Starfish (Asteroidea) and sea urchins (Echinoidea) are the most familiar echinoderm classes. Other classes include the brittle stars (Ophiuroidea), crinoids (Crinoidea) and blastoids (Blastoidea). Crinoids are more familiar as fossils than as modern organisms, and blastoids are known only from the fossil record.

Echinoderm skeletons consist of numerous calcite crystals. This might seem remarkable to mineral collectors who do not generally associate crystals with living things. Shells of snails and clams are made of calcite or aragonite. These, however, are composed of numerous grains too small to discern. The bones and teeth of vertebrates contain the mineral hydroxylapatite.

Each Piece a Single Crystal, Almost

Sea urchins are spiny creatures that may be encountered when bathing on a rocky beach or when snorkeling around a coral reef. It is best not to step on them, as the numerous spines can produce wounds that do not heal easily. Sand dollars are flat sea urchins with short spines that create the appearance of a hairy coat; they are often found in shallow water along sandy beaches. The main body skeleton, or *corona*, of the sea urchin is often found on those same rocky or sandy beaches devoid of its spiny covering.

The empty coronas are fragile and brittle; thus, whole specimens are only rarely found. Interestingly, close examination of corona fragments often reveals that they are not simply broken, but that there are visible *sutures* along which the shell comes apart. The corona of a sea urchin is composed of numerous individual plates that are located within the creature's skin and are held together by connective tissues. When these tissues decay, the plates separate quite easily from one another. The individual plates act as a single crystal, but have recently been discovered to be numerous individual crystals that are highly aligned crystallographically. The crystals are separated by a small amount of organic matter (Smith, 1989).

Holes for Strength

Under magnification it becomes obvious that between the crystals, the plates are full of interconnected pores, and as little as 50 percent of the bulk of the plate is actually calcite. In the living organism, these pores are filled with fluid and with living tissue. These pores also serve, paradoxically, to strengthen the skeletal plates (Nichols, 1969). Were the plates a single, solid crystal, they would break quite easily along cleavage directions. Since, however, the plates are made up of numerous tiny crystals with pores between them, it is more difficult for a crack to grow than it would be were the crack occurring across a solid crystal of the same thickness. The pores thus strengthen the crystal plate, which when broken shows rough and irregular fracture surfaces. The fluids and tissues that occupy the pores in the plates further help to strengthen the plates, as the thin organic films that separate the individual crystal elements are also presumed to do.

When echinoderm skeletons are fossilized, calcite cement typically fills the pores in the skeletal plates. The cement is aligned in accordance with the orientation of the calcite already in each plate. When the process of fossilization is complete, echinoderm skeletal elements behave like other calcite crystals and break along the usual cleavage directions. Many Paleozoic limestone deposits in the midwestern United States have abundant crinoid plates. Crinoid plates are immediately recognizable in the field because of their flat cleavage surfaces. On rare occasions, well-formed calcite crystals are even found as oriented overgrowths on echinoderm plates!

Crystal Spines

Each of the spines of the sea urchin is also made up of a single calcite crystal aligned in a single direction. Some of these spines can be considerable in size. The *c*-axes of these spine-forming crystals are oriented parallel to the length of the spine (Raup, 1966). This is easy to see in fossilized (but not in unfossilized) sea urchin spines because the cleavage surfaces are all equally inclined relative to the long dimension. This orientation may have evolved to optimize

Sea Urchins and other Echinoderms

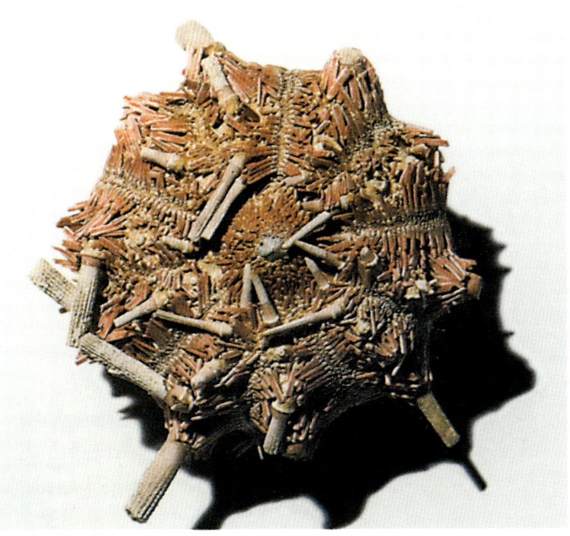

Armor and spines made of calcite

Above left: *A 4 cm widecommon sea urchin that was found on the shores of the Mediterranean Sea. This specimen has retained some of its spines.*

Above right: *A fossilized Cretaceous sea urchin found in France. The plates that form the corona of the sea urchin are clearly visible.*

Middle left: *A recent sea urchin corona without its spiny covering*

Middle right: *The petrified corona of Plegiocidaris coronata found in Nattheim (Germany).*

Below: *Two Jurassic sea urchin spines; the smaller spine belongs to the species Hemicidaris, the larger one to the species Rhabdocidaris. The right one is broken with a clean rhombohedral cleavage face. The enlarged detail on the right shows that the cleavage surface is inclined with regard to the long axis of the spine (the c-axis!)*

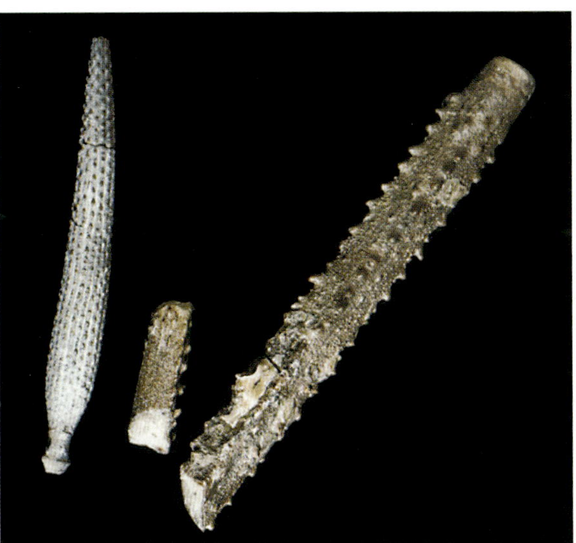

All specimens and photos Rupert Hochleitner

A crinoid, or sea lily, is attached to the sea floor by a stem of stacked disk-shaped plates.

Top and middle: *Calcite crystals overgrow the stem plates of a crinoid. The pinkish objects inside the crystals are individual discs or plates that extend from the stem. The crystals are about 1.5 mm long.*

Right: *Even the calcite crystals that overgrow a croinoid stem sometimes cleave. The cleavage plane passes through the crystal and the enclosed crinoid ossicle (stem), demonstrating the oriented nature of the overgrowth.*

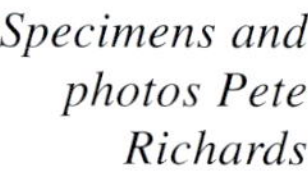

Specimens and photos Pete Richards

the strength of the spines because it maximizes the area of the shortest cleavage direction through the spine. The shortest cleavage direction is the most likely direction in which a break will occur.

Crystallography of the Sea Urchin

The orientations of the *c*-axes of the plates that form the corona can be different from one another. In fact the corona of a sea urchin is made up of five ambulacral sections, through which the nerves and blood vessels run, these alternate between five interambulacral sections. The five-fold symmetry of the sea urchin is obvious.

There are sea urchins in which the *c*-axes of the crystals of both types of plates point radially outward. In other individuals, the *c*-axes of the ambulacral sections are oriented toward the creature's exterior, while the interambulacral sections are oriented in the opposite direction, tangential to the curve of the sea urchin shell. Still other sea urchins only have plates in which the *c*-axis is tangential. Some of the families within the class of sea urchins exhibit plates for which the *c*-axes are inclined, in which case the angle of inclination changes during the course of shell growth.

The crystallography of the sea urchin thus varies. As a rule, each family within the class Echinoidea has roughly the same crystallography. Families that are taxonomically distinct also have clear crystallographic differences.

Brittle Star Optics

While the crystallography of echinoderms may seem to be of only academic interest, at least one group of applied researchers would most certainly debate the point.

Scientists from Lucent Technologies' Bell Labs, the Weizmann Institute of Science in Israel and the Natural History Museum of Los Angeles County have found that the calcite crystals making up the spines of brittle stars act both as the organism's armor and as the receptors for a sort of compound eye. Publishing their findings in the August 2001 issue of *Nature*, the group expects their discovery to lead to improving the optical technologies that are used in telecommunications as well as other industries.

Every serious calcite collector clearly needs at least one echinoderm to complete his collection.

References

Aizenberg, Tkachenko, Weiner, Addadi & Hendler (2001) Calcitic microlenses as part of the photoreceptor system in brittlestars. *Nature* 412: 819 - 822

Bartholinus (1669) *Experimenta Crystalli Islandici Disdiaclastici*. Copenhagen, D. Paulli, 60 p.

Bloss (1982) *The spindle stage: principles and practice*. Cambridge University Press, Cambridge, England. 340 p.

Bloss (1999) *Optical crystallography*. Mineralogical Society of America, Washington, D.C. 239 p.

Bostwick (1982) A Brief Review of Mineral Fluorescence at Franklin and Sterling Hill. *Rocks & Minerals* 57 : 196-201.

Brewster, D. (1818) On the laws of polarization and double refraction in regularly crystallized bodies. *Philoso-phical Transaction of the Royal Society* 108: 199-273.

Diderot and d'Alembert (1765) *Encyclopedia - Dictionnaire Raisonné des Sciences, des Arts et des Métiers*. Paris, France.

Duthaler & Weiss (1999) *Mineralien Reinigen und Aufbewahren*. Christian Weise Verlag, Munich, Germany. 49-55.

Eriksson (1995) Only a Calcite. *Rocks & Minerals* 70: 217-230.

Francis (1982) Fluorite and Associated Minerals from Southern Illinois. *Rocks & Minerals* 57: 63-71.

Gaines, Skinner, Foord, Mason & Rosenzweig (1997) *Dana's New Mineralogy: the system of mineralogy of James Dwight Dana and Edward Salisbury Dana*, 8th ed. John Wiley & Sons, New York.

Gebhard (1999) *Tsumeb: a unique mineral locality*. GG Publishing, Grossenseifen, Germany.

Goldschmidt (1913) *Atlas der Krystallformen*, vol 2. Carl Winters Universitätsbuchhandlung, Heidelberg, Germany.

Grant & Wilson (2001) Famous mineral localities: Dal'negorsk, Primorskiy Kray, Russia. *Mineralogical Record* 32: 3-30.

Greene (2001) TAG Speleothemes. *The Winged Messenger, Upper Cumberland Grotto's Newsletter*

Grundmann (1998) Fundorte von Calcite – Die Top Ten der Weltrangliste: Indien-West. *extraLapis No. 14 Calcit*: 64. Christian Weise Verlag, München.

Günther (1982) Polyedrische Quarzdrusen aus Brasilien. *Aufschluss* 33: 147-151.

Hansen (1984) Cleaning Delicate Minerals. *Mineralogical Record* 15, 2: 103.

Haüy (1782) Observation sur La Physique, sur l'Histoire Naturelle et les Arts. *Journal de Physique* also known as *Physique de Rozier*. Paris, France.

Haüy (1801) *Traité de Minéralogie*. Paris, France.

Hecht & Zajac (1979) *Optics*. 4th Edition, Addison-Wesley Publishing Company, Reading, Massachusetts, 565 p.

Hill & Forti (1986) *Cave Minerals of the World*. National Speleological Society, Huntsville Alabama. 238 p.

Hooykaas (1955) Les débuts de la théorie cristallographique de R.J. Haüy d'aprés les documents originaux. *Revue Hist. Sc.* T.8 n°4

Huizing & Russell (1986) Indiana Minerals: A Locality Index. *Rocks & Minerals* 61: 136-151.

Hurlbut & Francis (1984) Notes and New Techniques, An Extraordinary Calcite Gemstone. *Gems & Gemology* winter issue.

Huygens (1660) *Traité de la Lumière*. Leyden, Holland

Kelly (1994) *The Red Hills. Iron mines of West Cumberland*. Red Earth Publications, Ulverston, Cumbria.

Kile (2003) The petrographic microscope: Evolution of a Mineralogical Research Instrument. *Mineralogical Record*, in press.

Historic postcard showing a Joplin Missouri mining scene at the turn of the 20th century. Günter Grundmann collection

Köhler & Leithmeier (1933) Das Verhalten des Kalkapates im ultravioletten Licht. *Zentralblatt für Mineralogie, Geologie und Paläontologie 12: 401-411*

Kolodny, Luz, Sanders & Clemens (1996) Dinosaur bones: fossils or pseudomorphs? The pitfalls of physiology reconstruction from apatitic fossils. *Palaeogeography, palaeoclimatology, palaeoecology* 126: 1/2; 161-171.

Kostov (1968) *Mineralogy*. Oliver and Boyd, London.

Kristjansson (2002) Iceland spar: The Helgustadir calcite locality and its influence on the development of science. *The Journal of Geoscience Education* 50, 4: 419-427.

Lasmanis (1989) Galena from Mississippi Valley-Type Deposits. *Rocks & Minerals* 64: 10-34.

Lieber (1969) *Mineralogie in Stichworten*, Kiel, Germany.

Lillie (1988) Minerals of the Harris Creek Fluorspar District, Hardin County, Illinois. *Rocks & Minerals* 63: 210-226.

Malus (1808) Sur une propriete de la lumiere reflechie. *Bulletin de Societe Philomathique* v. 1, 266-269.

Meter (1989) *Auswanderung von Hunsrück nach Brasilien*. Verein für Heimatkunde Nonnweiler, 96 p.

Miers (1889) Calcites from the neighbourhood of Egremont, Cumberland. *Mineralogical Magazine* 8: 149-153.

Modreski & Aumente-Modreski (1996) Fluorescent Minerals - A Review. Rocks & Minerals 71: 14-22.

Moroshkin & Frishman (2001) Dal'negorsk: Notes on Mineralogy. *Mineralogical Almanac* 4: 1-128.

Mutschler (1954) The luminescent minerals of Franklin, New Jersey. *Rocks and Minerals* 29: 482-485.

Nesse (1991) *An introduction to optical mineralogy*. 2nd ed. Oxford University Press, New York. 335 p.

Nichols (1969) *Echinoderms*, 4th Edition. Hutchinson, London.

Nicol (1829) On a method of so far increasing the divergency of the two rays in calcareous-spar, that only one may be seen at a time. *Edinburgh New Philosophical Journal* 6: 83-84.

Orrell (1993) Mining California calcite crystals for the optical ring sight. *California Geology,* March/April, 45-49.

Ottens (2003) Indian Zeolithes and Related Species. *Mineralogical Record* 34:1

Otto & Senf (1992) Reinigen, Präparieren und Konservieren von Mineralstufen, *Lapis* 17, 12: 24.

Palache (1943) Calcite, an angle table and critical list. *Harvard University Department of Mineralogy and Petrology publication* 259.

Palache, Berman & Frondel (1951) *The System of Mineralogy of James Dwight Dana and Edward Salisbury Dana*, 7th ed. John Wiley & Sons, New York.

Panczner (1987) *Minerals of Mexico*. Van Nostrand Reinhold, New York.

Pennypacker (1896) Calcite. *Mineral Collector* 3: 154-155.

Presmyk (1997) Brushy Creek Mine, Reynolds County, Missouri. *Rocks & Minerals* 72: 370-377.

Raup (1966) The endoskeleton. In R.A. Boolootian, ed. *Physiology of Echinodermata*. Wiley Interscience, New York. 379-395

Reeder (1983) *Reviews in Mineralogy*. Vol. 11. Carbonates: Mineralogy and Chemistry. Mineralogical Society of America.

Richards (1999) The Four Twin-laws of Calcite, *Rocks & Minerals* 74: 308-317.

Romé de l'Isle (1783) *Cristallographie ou Déscription des formes propres a tous le corps du règné minéral.* Paris, France.

Rudler (1905) *A handbook to a collection of the minerals of the British Islands, mostly selected from the Ludlam Collection, in the Museum of Practical Geology, Jermyn Street, London, S.W. HMSO*, London. 241p.

Schröcke & Weiner (1981) *Mineralogie*. De Gruiter. 952 p.

Shaffer (1981) Possibility of Mississippi Valley-Type Mineral Deposits in Indiana. Dept. of Natural Resources, *Geological Survey Special Report* 21.

Sinkankas (1964) *Mineralogy for Amateurs*. Van Nostrand, Princeton, New Jersey.

Smirnov (1976) *Geology of Mineral Deposits*. Moscow: MIR Publications.

Smith (1989) Biomineralization in Echinoderms. In J.G. Carter, ed., Skeletal Biomineralization: Patterns, processes, and evolutionary trends. *AGU Short Course in Geology* 5 No. 2: 117-147.

Strunz (1967) *Klockmanns Lehrbuch der Mineralogie*. 554 p.

Waller (1980) A Rust Removal Method for Mineral Specimens, Notes for Collectors. *Mineralogical Record* 11: 109.

Whitlock (1909) Calcites of New York. *New York State Museum 63rd Annual Report* 4: 1-190.

Wilke (1978): Die Erzgänge von Andreasberg im Rahmen des Mittelharz-Ganggebietes. *Beihefte Geologisches Jahrbuch*, 7: Hannover, 228 p.

Wilson & Dyl (1992) Michigan Copper Country. *Mineralogical Record* 23: 1-104.

Wood, E.A. (1964) *Crystals and light*. D. Van Nostrand, Princeton, New Jersey, 160 p.

Authors' Addresses

Michael P. Cooper
Nottingham City Museums and Galleries
Brewhouse Yard, Castle Boulevard
Nottingham NG71FB
United Kingdom
minerals@mcooper.demon.co.uk

Dr. Rudolf Duthaler
Girenhaldenweg 17
CH-4126 Bettingen, Switzerland
rudolf.duthaler@pharma.novartis.com

Stanley J. Dyl
A. E. Seaman Mineral Museum
516 EERC Building; 1400 Townsend Dr.
Houghton, MI 49931-1295 USA
sjdyl@mtu.edu

Dr. Günter Grundmann
Technische Universität München
Arcisstrasse 21
D-85290 München, Germany
guenter.grundmann@geo.tum.de

Dr. Mickey Gunter
Department of Geological Sciences
University of Idaho
Moscow, ID 83844-3022 USA
mgunter@uidaho.edu

Dr. Rupert Hochleitner
Mineralogische Staatssammlung
Theresienstrasse 41
D-80333 Munich, Germany
rh.minstaatsslg@lrz.uni-muenchen.de

Terry Huizing
5341 Thrasher Drive
Cincinnati, OH 45247 USA
tehuizing@fuse.net

Dr. Werner Lieber
Baden-Badener-Strasse 3
D-69126 Heidelberg, Germany

Dr. Guanghua Liu
Französische Allee 24
D-72072 Tübingen, Germany
aaamin@web.de

Bertold Ottens
Klingenbrunn Bahnhof 24
D-94518 Spiegelau, Germany
ottens-mineralien@t-online.de

Dr. R. Peter Richards
154 Morgan Street
Oberlin, Ohio 44074 USA
peterichards@oberlin.net

Marc L. Wilson
Carnegie Museum of Natural History,
4400 Forbes Avenue
Pittsburgh, PA 15213-4080 USA
wilsonm@carnegiemuseums.org

Reinhard Balzer
Eichendorffstrasse 7
D-35039 Marburg/Lahn, Germany
rbalzer-marburg@t-online.de

Sarah Bronko
c/o Lapis International, LLC
Post Office Box 263
East Hampton, CT 06424
sbronko@hamilton.edu

Michael Gray
Post Office Box 727
Missoula, MT 59806-0727 USA
michael@gray.name

Dr. Lydie Touret
Musée de Minéralogie - Ecole des Mines
60 Boulevard St. Michel
F-75272 Paris Cedex 06 France
touret@musee.ensmp.fr